A SURVEY OF MATRIX THEORY AND MATRIX INEQUALITIES

MARVIN MARCUS
Professor of Computer Science
University of California, Santa Barbara

HENRYK MINC
Professor of Mathematics
University of California, Santa Barbara

DOVER PUBLICATIONS, Inc., *New York*

To Arlen and Catherine

Published in Canada by General Publishing Company, Ltd., 30 Lesmill Road, Don Mills, Toronto, Ontario.
Published in the United Kingdom by Constable and Company, Ltd., 3 The Lanchesters, 162–164 Fulham Palace Road, London W6 9ER.

This Dover edition, first published in 1992, is an unabridged, unaltered republication of the corrected (1969) printing of the work first published by Prindle, Weber & Schmidt, Boston, 1964, as Volume Fourteen of "The Prindle, Weber & Schmidt Complementary Series in Mathematics."

Manufactured in the United States of America
Dover Publications, Inc., 31 East 2nd Street, Mineola, N.Y. 11501

Library of Congress Cataloging-in-Publication Data

Marcus, Marvin, 1927–
    A survey of matrix theory and matrix inequalities / Marvin Marcus, Henryk Minc.
        p.   cm.
    Originally published: Boston : Prindle, Weber & Schmidt, 1969.
    Includes bibliographical references.
    ISBN 0-486-67102-X (pbk.)
    1. Matrices.   I. Minc, Henryk.   II. Title.
QA188.M36   1992
512.9′434—dc20   91-39704
                                                                          CIP

Preface

MODERN MATRIX THEORY is a subject that has lately gained considerable popularity both as a suitable topic for undergraduate students and as an interesting area for mathematical research. Apart from aesthetic reasons this is a result of the widespread use of matrices in both pure mathematics and the physical and social sciences.

In this book we have tried within relatively few pages to provide a survey of a substantial part of the field. The first chapter starts with the assumption that the reader has never seen a matrix before, and then leads within a short space to many ideas which are of current research interest. In order to accomplish this within limited space, certain proofs that can be found in any of a very considerable list of books have been left out. However, we have included many of the proofs of classical theorems as well as a substantial number of proofs of results that are in the current research literature. We have found that a subset of the material covered in Chapter I is ample for a one-year course at the third- or fourth-year undergraduate level. Of course, the instructor must provide some proofs but this is in accordance with our own personal preference not to regurgitate "definition-theorem-proof" from the printed page. There are many ideas covered in Chapter I that are not standard but reflect our own prejudices: e.g., Kronecker products, compound and induced matrices, quadratic relations, permanents, incidence matrices, (v, k, λ) configurations, generalizations of commutativity, property L.

The second chapter begins with a survey of elementary properties of convex sets and polyhedrons and goes through a proof of the Birkhoff theorem on doubly stochastic matrices. From this we go on to the properties of convex functions and a list of classical inequalities. This material is then put together to yield many of the interesting matrix inequalities of Weyl, Fan, Kantorovich, and others. The treatment is along the lines developed by these authors and their successors and many of the proofs are included. This chapter contains an account of the classical Perron-Frobenius-Wielandt theory of indecomposable nonnegative matrices and ends with some very recent results on stochastic matrices.

The third chapter is concerned with a variety of results on the localization of the characteristic roots of a matrix in terms of simple functions of its entries or of entries of a related matrix. The presentation is essentially in historical order, and out of the vast number of results in this field we have tried to cull those that seemed to be interesting or to have some hope of being useful.

In general it is hoped that this book will prove valuable in some respect to anyone who has a question about matrices or matrix inequalities.

MARVIN MARCUS
HENRYK MINC

The present edition is unaltered except for minor typographical corrections.

M. M.
H. M.
August, 1969

Contents

I. SURVEY OF MATRIX THEORY

vii

Special Notation

$\mathbf{R}$—real numbers.

$\mathbf{C}$—complex numbers.

$\mathbf{I}$—integers.

$\mathbf{P}$—nonnegative real numbers.

$\mathbf{F}[\lambda]$—polynomial in λ with coefficients in $\mathbf{F}$.

$M_{m,n}(\mathbf{F})$—$m \times n$ matrices with entries in $\mathbf{F}$.

$M_n(\mathbf{F})$—n-square matrices with entries in $\mathbf{F}$.

$A_{(i)}$—ith row of A.

$A^{(j)}$—jth column of A.

$\mathbf{F}^k$—set of k-tuples of numbers from F.

$0_{m,n}$—$m \times n$ zero matrix.

I_n—n-square identity matrix.

0_n—n-tuple of zeros.

A^T—transpose of A.

A^*—conjugate transpose of A.

$\overline{A}$—conjugate of A.

$A \dotplus B$ or $\overset{k}{\underset{i=1}{\Sigma}} \cdot A_i$—direct sum.

$A \otimes B$—Kronecker product.

$Q_{k,n}$—totality of strictly increasing sequences of k integers chosen from $1, \cdots, n$.

$G_{k,n}$—totality of nondecreasing sequences of k integers chosen from $1, \cdots, n$.

$D_{k,n}$—totality of sequences of k distinct integers chosen from $1, \cdots, n$.

$S_{k,n}$—totality of sequences of k integers chosen from $1, \cdots, n$.

$E_k(a)$—kth elementary symmetric function of the numbers
   $a = (a_1, \cdots, a_n)$.
$A[\alpha|\beta]$, $A[\alpha|\beta)$, $A(\alpha|\beta]$, $A(\alpha|\beta)$—submatrices of A using rows numbered α
   and columns numbered β, rows α and columns complementary to β,
   etc. Here α and β are sequences of integers.
S_n—symmetric group of degree n.
$d(A)$—determinant of A.
$C_r(A)$—rth compound matrix of A.
$u_1 \wedge \cdots \wedge u_r$—skew-symmetric product of the vectors $u_1, \cdots, u_r$.
$E_k(A)$—sum of all k-square principal subdeterminants of A.
tr (A)—trace of A.
$p(A)$—permanent of A.
$P_r(A)$—rth induced matrix of A.
$u_1 \cdots u_r$—symmetric product of the vectors $u_1, \cdots, u_r$.
$\lambda_i(A)$—a characteristic root of A.
$\rho(A)$—rank of A.
e_i—the n-tuple with 1 as ith coordinate, 0 otherwise.
$\langle x_1, \cdots, x_k \rangle$—space spanned by $x_1, \cdots, x_k$.
$\eta(A)$—nullity of A.
$I_{(i),(j)}$—interchange rows i and j.
$II_{(i)+c(j)}$—add c times row j to row i.
$III_{c(i)}$—multiply row i by c.
$\left.\begin{array}{l} E_{(i),(j)} \\ E_{(i)+c(j)} \\ E_{c(i)} \end{array}\right\}$ elementary matrices (row).
$I^{(i),(j)}$—interchange columns i and j.
$II^{(i)+c(j)}$—add c times column j to column i.
$III^{c(i)}$—multiply column i by c.
$\left.\begin{array}{l} E^{(i),(j)} \\ E^{(i)+c(j)} \\ E^{c(i)} \end{array}\right\}$ elementary matrices (column).
$a \mid b$—a divides b.
$C(p(\lambda))$—companion matrix of $p(\lambda)$.
$H(p(\lambda))$—hypercompanion matrix of $p(\lambda)$.
diag $(a_1, \cdots, a_n)$—the diagonal matrix with $a_1, \cdots, a_n$ down the main
   diagonal.

CHAPTER II

U—(in Chapter II) either of the sets $\mathbf{R}^n$ or $M_{m,n}(\mathbf{R})$.
$\langle x_1, \cdots, x_k \rangle^{\perp}$—the orthogonal complement of the vector space spanned
   by $x_1, \cdots, x_k$.
D_n—set of all real n-square matrices with row and column sums equal to 1.

Ω_n—set of all n-square doubly stochastic matrices.

$H(X)$—convex hull of X.

$A * B$—Hadamard product (the i,j entry is $a_{ij}b_{ij}$).

CHAPTER III

In Sections **1.2, 1.3, 1.4,** and **1.6,** the following special notation is used: for $A \in M_n(\mathbf{C})$ let $B = (A + A^*)/2$, $C = (A - A^*)/2i$; let $\lambda_1, \cdots, \lambda_n$ ($|\lambda_1| \geq \cdots \geq |\lambda_n|$), $\mu_1 \geq \cdots \geq \mu_n$, $\nu_1 \geq \cdots \geq \nu_n$ be the characteristic roots of A, B, and C respectively; let $g = \max_{i,j} |a_{ij}|$, $g' = \max_{i,j} |b_{ij}|$, $g'' = \max_{i,j} |c_{ij}|$.

In Sections **1.6, 2.1, 2.2, 2.4,** and **2.5;** if $A \in M_n(\mathbf{C})$, then R_i denotes the sum of the absolute values of the entries in row i; T_i is the sum of the absolute values of the entries in column i; R is the largest of the R_i; T is the largest of the T_i; $P_i = R_i - |a_{ii}|$; $Q_i = T_i - |a_{ii}|$.

$s(A)$—the spread of A.

$||A||$—Euclidean norm of $A \in M_{m,n}(\mathbf{C})$ ($||A||^2 = \sum\limits_{i,j=1}^{m,n} |a_{ij}|^2$).

The Numbering System

WITHIN THE TEXT of a chapter a reference to a section in that chapter will give only the number of the section. A reference to a section in some other chapter will be preceded by the number of that chapter. Thus, if in the text of Chapter II we wish to refer to Section **2.12** of Chapter I we write "(see **I.2.12**)." Exhibited formulas are referred to by the section number followed by the formula number in parentheses. Thus, if in the text of Chapter III we wish to refer to formula (2) of Section **3.4** of Chapter I, we write "[see **I.3.4(2)**]."

Survey of Matrix Theory

I

1 Introductory Concepts

1.1 Matrices and vectors

In this book we will be dealing with the following sets:

(i) The field of real numbers **R**.
(ii) The field of complex numbers **C**.
(iii) The ring of all integers (positive, negative, and zero) **I**.
(iv) The set of all nonnegative real numbers **P**.
(v) The ring of polynomials in some indeterminate ("variable") λ with real (complex) coefficients $\mathbf{R}[\lambda]$ ($\mathbf{C}[\lambda]$).

Let **F** be any one of these sets and suppose that $a_{11}, a_{12}, a_{13}, \cdots, a_{mn}$ is a collection of mn elements in **F**. The rectangular array of these elements

$$
\begin{pmatrix}
a_{11} & a_{12} & \cdots & a_{1n} \\
a_{21} & a_{22} & & a_{2n} \\
\cdot & & & \\
\cdot & & & \\
\cdot & & & \\
a_{m1} & a_{m2} & & a_{mn}
\end{pmatrix}
\tag{1}
$$

consisting of m rows and n columns is called an $m \times n$ *matrix*. We write $A = (a_{ij})$. A more formal but less suggestive definition is this: a matrix A

is a function on the set of pairs of integers (i,j), $1 \leq i \leq m$, $1 \leq j \leq n$, with values in $\mathbf{F}$ in which a_{ij} designates the value of A at the pair (i,j). The array **1.1(1)** is then just a convenient way of exhibiting the range of the function A. The element a_{ij} is called the (i,j) *entry* of A. The ith *row* of A is the sequence $a_{i1}, a_{i2}, \cdots, a_{in}$, and the jth *column* is the sequence $a_{1j}, a_{2j}, \cdots, a_{mj}$. Since the a_{ij} are in $\mathbf{F}$, we say that A is an $m \times n$ matrix *over* $\mathbf{F}$. The totality of such matrices will be designated by $M_{m,n}(\mathbf{F})$. In case $n = m$, we say that A is an *n-square* matrix. Denote by $M_n(\mathbf{F})$ the set of all n-square matrices over $\mathbf{F}$. A *row (column) vector* over $\mathbf{F}$ is just an element of $M_{1,n}(\mathbf{F})$ $(M_{m,1}(\mathbf{F}))$. If $A \in M_{m,n}(\mathbf{F})$, then $A_{(i)} = (a_{i1}, \cdots, a_{in}) \in M_{1,n}(\mathbf{F})$, $(i = 1, \cdots, m)$, designates the ith row vector of A. Similarly, the column vector $A^{(j)} \in M_{m,1}(\mathbf{F})$ designates the $m \times 1$ matrix whose (i,j) entry is a_{ij}, $(i = 1, \cdots, m)$. The (i,j) entry of A is sometimes designated by A_{ij} also. In general, a *k-vector v over* $\mathbf{F}$ is just an ordered k-tuple of elements of $\mathbf{F}$, $(a_1, a_2, \cdots, a_k)$; a_i is called the ith *coordinate* of v. The k-vector every one of whose coordinates is 0 will be denoted by 0 also. In case it is necessary to designate its size, we write 0_k. The totality of k-vectors over $\mathbf{F}$ is denoted by $\mathbf{F}^k$. In case $\mathbf{F} = \mathbf{C}$, $v = (a_1, a_2, \cdots, a_k) \in \mathbf{C}^k$, then $\bar{v}$ denotes the vector $(\bar{a}_1, \bar{a}_2, \cdots, \bar{a}_k)$. Two special matrices are the n-square *identity* matrix $I_n \in M_n(\mathbf{F})$ and the $m \times n$ *zero* matrix $0_{m,n} \in M_{m,n}(\mathbf{F})$. The (i,j) entry of I_n is 0 unless $i = j$, in which case it is 1. That is, the (i,j) entry of I_n is the Kronecker delta, $\delta_{ij} = 0$ if $i \neq j$, $\delta_{ij} = 1$ if $i = j$. The (i,j) entry of $0_{m,n}$ is 0 for every i and j.

1.2   Matrix operations

Three standard *operations* for matrices over $\mathbf{F}$ are defined below.

1.2.1   *Multiplication of $A = (a_{ij}) \in M_{m,n}(\mathbf{F})$ by a scalar*, i.e., an element $c \in \mathbf{F}$. The product is defined as the matrix in $M_{m,n}(\mathbf{F})$ whose (i,j) entry is ca_{ij} and is designated by cA.

1.2.2   The *sum* of two matrices $A = (a_{ij})$ and $B = (b_{ij})$ in $M_{m,n}(\mathbf{F})$ is the matrix $C = (c_{ij}) \in M_{m,n}(\mathbf{F})$ whose (i,j) entry is $c_{ij} = a_{ij} + b_{ij}$. Observe that the matrices A and B must be the same size in order that the sum be defined.

1.2.3   The *product* of two matrices $A = (a_{ij}) \in M_{m,k}(\mathbf{F})$ and $B = (b_{ij}) \in M_{k,n}(\mathbf{F})$ is the matrix $C = (c_{ij}) \in M_{m,n}(\mathbf{F})$ whose (i,j) entry is

$$c_{ij} = \sum_{t=1}^{k} a_{it} b_{tj}, \qquad (1 \leq i \leq m, 1 \leq j \leq n). \tag{1}$$

We write $C = AB$, using simply proximity to indicate the product. Observe that in order for the product $C = AB$ to be defined the number of columns of A must be the same as the number of rows of B. If $A \in M_n(\mathbf{F})$, we use the notation A^2 to designate the product AA; $A^3 = AA^2$, and in general $A^r = AA^{r-1}$ for any positive integer r. The matrix A^r is called the rth *power* of A for obvious reasons.

1.2.4   The rules connecting these operations are:

(i)   *Commutativity*: $A + B = B + A$. However, $AB \neq BA$ in general. If $AB = BA$, then A and B are said to *commute*.

(ii)   *Associativity*: $A + (B + C) = (A + B) + C$; $A(BC) = (AB)C$. The common values of these sums (products) are written without parentheses: $A + B + C$ (ABC).

(iii)   *Distributivity*: $A(B + C) = AB + AC$, $(B + C)A = BA + CA$.

Of course all the matrices in (i), (ii), and (iii) must be of appropriate sizes for the indicated operations to be defined.

1.3 Inverse

If $A \in M_n(\mathbf{F})$, it is sometimes the case that there exists a matrix $B \in M_n(\mathbf{F})$ such that $BA = AB = I_n$. If this happens, A is said to be *nonsingular*. If no such B exists, then A is called *singular*. If A is nonsingular, then B is uniquely determined by the equation $AB = I_n$ or by $BA = I_n$. The matrix B is called the *inverse* of A, and the notation A^{-1} is used to designate B. If r is a positive integer, then A^{-r} denotes the matrix $(A^{-1})^r$; $A^0 = I_n$. The usual laws of exponents $A^{p+q} = A^p A^q$ and $(A^p)^q = A^{pq}$ work for any A if p and q are positive integers. If A is nonsingular, these laws work for positive, zero, and negative integer exponents p and q.

1.4 Matrix and vector operations

Operations involving k-vectors and matrices over $\mathbf{F}$ are also defined.

1.4.1   The *sum* of two vectors $u = (a_1, \cdots, a_k)$ and $v = (b_1, \cdots, b_k)$ is the vector $w = u + v = (a_1 + b_1, \cdots, a_k + b_k)$.

1.4.2   If $c \in \mathbf{F}$ and $u = (a_1, \cdots, a_k)$, then the *scalar product* $cu = uc$ is the vector $(ca_1, \cdots, ca_k)$.

1.4.3   If $A = (a_{ij}) \in M_{m,k}(\mathbf{F})$ and $u = (c_1, \cdots, c_k)$, then the *matrix-vector* product Au is defined as the vector $v = (b_1, \cdots, b_m)$ for which

$$b_j = \sum_{t=1}^{k} a_{jt} c_t, \qquad (j = 1, \cdots, m). \tag{1}$$

Clearly $A(u + v) = Au + Av$ and $(A + B)u = Au + Bu$ for any vectors u and v in $\mathbf{F}^k$ and any matrices A and B in $M_{m,k}(\mathbf{F})$.

1.4.4 We can express the equation **1.4(1)** in another way in terms of column vectors: if $A \in M_{m,k}(\mathbf{F})$ and $u = (c_1, \cdots, c_k)$, then

$$v = \sum_{t=1}^{k} c_t A^{(t)},$$

for the jth coordinate of the right side of this is $\sum_{t=1}^{k} a_{jt}c_t$.

Similarly, if $w = (d_1, \cdots, d_m) \in M_{1,m}(\mathbf{F})$ and $A = (a_{ij}) \in M_{m,k}(\mathbf{F})$, then

$$wA = \sum_{s=1}^{m} d_s A_{(s)}.$$

We can also express the definition **1.2.3** for the product of two matrices in terms of column and row vectors of the matrices involved. For, in the notation of **1.2.3**, if $C = AB$, then

$$C^{(j)} = \sum_{t=1}^{k} b_{tj} A^{(t)}, \qquad (j = 1, \cdots, n),$$

and   (2)

$$C_{(i)} = \sum_{t=1}^{k} a_{it} B_{(t)}, \qquad (i = 1, \cdots, m).$$

The formulas can also be viewed as matrix-vector products, namely

$$C^{(j)} = AB^{(j)}, \qquad (j = 1, \cdots, n),$$

and   (3)

$$C_{(i)} = A_{(i)}B, \qquad (i = 1, \cdots, m).$$

1.5 Examples

1.5.1 If $A \in M_{m,k}(\mathbf{F})$, then $A0_{k,n} = 0_{m,n}$. For, setting $B = 0_{k,n}$ in **1.2(1)** results in $a_{it}b_{tj}$ being 0 for every i, j, t.

1.5.2 If $A \in M_{m,k}(\mathbf{F})$, then $I_m A = A = AI_k$ and $A + 0_{m,k} = A$. For example, if $C = AI_k$, then $c_{ij} = \sum_{t=1}^{k} a_{it}\delta_{tj} = a_{ij}$.

1.5.3 If $A \in M_n(\mathbf{F})$ is the matrix whose $(i, i+1)$ entry is 1, $(i = 1, \cdots, n-1)$, with all other entries 0, then $A^n = 0_{n,n}$. To see this we show by induction that the matrix A^r has 1 as its $(i, i+r)$ entry, $(i = 1, \cdots, n-r)$, and all other entries 0. For $r = 1$, this is the case by the definition of A. Assume A^{r-1} has $\delta_{i+r-1,j}$ as its (i,j) entry. Set $C = A^r = AA^{r-1}$ and compute that

$$c_{ij} = \sum_{t=1}^{n} a_{it}\delta_{t+r-1,j}$$

$$= \sum_{t=1}^{n} \delta_{i+1,t}\delta_{t+r-1,j}$$

$$= \delta_{i+1+r-1,j} = \delta_{i+r,j}.$$

In other words, $c_{ij} = 0$ unless $j = i + r$, in which case it is 1. Of course if $r \geq n$, then $c_{ij} = 0$ for all i and j and we have the statement that $A^n = 0_{n,n}$.

Any matrix $A \in M_n(\mathbf{F})$ is called *nilpotent* if $A^r = 0_{n,n}$ for some positive integer r.

1.5.4   Let $A = (\delta_{i+1,j}) \in M_n(\mathbf{F})$ be the matrix discussed in **1.5.3** and let $e_1 = (1, 0, \cdots, 0)$ be the n-vector whose first coordinate is 1, the rest 0. Then if $v = Ae_1$, we compute directly from **1.4(1)** that the jth coordinate b_j of v is

$$b_j = \sum_{t=1}^{n} \delta_{j+1,t}\delta_{t1} = \delta_{j+1,1} = 0.$$

Thus $Ae_1 = 0$ even though A and e_1 are both different from 0.

1.6   Transpose

If $A \in M_{m,n}(\mathbf{F})$, then associated with A is the *transpose* of A, written A^T. This is the matrix in $M_{n,m}(\mathbf{F})$ whose (i,j) entry is the (j,i) entry of A, $(i = 1, \cdots, n; j = 1, \cdots, m)$. Various notations for the transpose are used: A'; A^t; $_tA$. When $\mathbf{F} = \mathbf{C}$, the set of complex numbers, then the matrix $A^* \in M_n(\mathbf{C})$ whose (i,j) entry is the complex conjugate of the (j,i) entry of A is called the *conjugate transpose* of A. The matrix A^* is also called the *adjoint* of A. For example,

$$\text{if} \quad A = \begin{pmatrix} 0 & i \\ i & 0 \end{pmatrix}, \quad \text{then} \quad A^* = -A = \begin{pmatrix} 0 & -i \\ -i & 0 \end{pmatrix}.$$

The matrix $\overline{A}$ whose (i,j) entry is $\overline{a}_{ij}$ is called the *conjugate* of A. Thus $A^* = \overline{A}^T$.

1.7   Direct sum and block multiplication

There are a couple of somewhat more complicated ways of putting matrices together to get new matrices. The first of these is the *direct sum*. If $A \in M_p(\mathbf{F})$ and $B \in M_q(\mathbf{F})$, then the direct sum of A and B, $A \dotplus B = C$, is the matrix in $M_{p+q}(\mathbf{F})$ whose entries are given by

$$\begin{aligned} c_{ij} &= a_{ij}, & (i, j = 1, \cdots, p), \\ c_{p+i,p+j} &= b_{ij}, & (i, j = 1, \cdots, q), \\ c_{ij} &= 0 \text{ otherwise.} \end{aligned}$$

Stated a little more graphically: the matrix $A \dotplus B$ is built by joining the upper left corner of B onto the lower right corner of A. Clearly this operation is noncommutative but it is associative. More formally define $A_1 \dotplus A_2 \dotplus A_3$ to be $(A_1 \dotplus A_2) \dotplus A_3$ and inductively define

$$\sum_{i=1}^{m}{}^{\cdot} A_i = \sum_{i=1}^{m-1}{}^{\cdot} A_i \dotplus A_m.$$

If $A_i = (a_i)$ is a 1-square matrix, $(i = 1, \cdots, m)$, then $\sum_{i=1}^{m}{}^{\cdot} A_i$ is sometimes written diag $(a_1, \cdots, a_m)$. If A_i is a p_i-square matrix, $(i = 1, \cdots, k)$, and $q = \sum_{i=1}^{k} p_i$, then $\sum_{i=1}^{k}{}^{\cdot} A_i$ is a q-square matrix. Suppose that A_i and B_i are p_i-square matrices, $(i = 1, \cdots, k)$. Then an important formula is

$$\sum_{i=1}^{k}{}^{\cdot} A_i \sum_{i=1}^{k}{}^{\cdot} B_i = \sum_{i=1}^{k}{}^{\cdot} A_i B_i. \tag{1}$$

This formula is actually an example of the more general principle of *block multiplication* of conformally partitioned matrices. Suppose $A \in M_{m,n}(\mathbf{F})$ and $B \in M_{n,r}(\mathbf{F})$. Let $m_1 + \cdots + m_d = m$, $n_1 + \cdots + n_c = n$, $r_1 + \cdots + r_f = r$, where m_i, n_j, r_k are all positive integers, $(i = 1, \cdots, d; j = 1, \cdots, c; k = 1, \cdots, f)$. Partition A and B into submatrices A_{ij}, B_{ke} as follows.

$$A = \begin{pmatrix} \overset{n_1}{\overbrace{A_{11}}} & \overset{n_2}{\overbrace{A_{12}}} & \cdots & \overset{n_c}{\overbrace{A_{1c}}} \\ A_{21} & A_{22} & \cdots & A_{2c} \\ \cdot & \cdot & & \cdot \\ \cdot & \cdot & & \cdot \\ \cdot & \cdot & & \cdot \\ A_{d1} & A_{d2} & \cdots & A_{dc} \end{pmatrix} \begin{matrix} \}m_1 \\ \}m_2 \\ \\ \\ \\ \}m_d \end{matrix},$$

$$B = \begin{pmatrix} \overset{r_1}{\overbrace{B_{11}}} & \overset{r_2}{\overbrace{B_{12}}} & \cdots & \overset{r_f}{\overbrace{B_{1f}}} \\ B_{21} & B_{22} & \cdots & B_{2f} \\ \cdot & \cdot & & \cdot \\ \cdot & \cdot & & \cdot \\ \cdot & \cdot & & \cdot \\ B_{c1} & B_{c2} & \cdots & B_{cf} \end{pmatrix} \begin{matrix} \}n_1 \\ \}n_2 \\ \\ \\ \\ \}n_c \end{matrix} . \tag{2}$$

A_{ij} is $m_i \times n_j$, B_{ke} is $n_k \times r_e$ and hence each product $A_{ij}B_{je}$ is defined and yields an $m_i \times r_e$ matrix as answer. The matrix product $C = AB \in M_{m,r}(\mathbf{F})$ is the matrix

$$C = \begin{pmatrix} \overset{r_1}{\overbrace{C_{11}}} & \overset{r_2}{\overbrace{C_{12}}} & \cdots & \overset{r_f}{\overbrace{C_{1f}}} \\ C_{21} & C_{22} & \cdots & C_{2f} \\ \cdot & \cdot & & \cdot \\ \cdot & \cdot & & \cdot \\ \cdot & \cdot & & \cdot \\ C_{d1} & C_{d2} & \vdots \cdots & C_{df} \end{pmatrix} \begin{matrix} \}m_1 \\ \}m_2 \\ \cdot \\ \cdot \\ \cdot \\ \}m_d \end{matrix}$$

in which C_{ie} is the $m_i \times r_e$ matrix

$$C_{ie} = \sum_{j=1}^{c} A_{ij}B_{je}, \qquad (i = 1, \cdots, d; e = 1, \cdots, f). \tag{3}$$

Thus when A and B are partitioned conformally into blocks, the product AB can be computed exactly as if the blocks were elements of $\mathbf{F}$. The only thing to be watched is the order of multiplying blocks together. In general, of course, $A_{ij}B_{je}$ is not the same as $B_{je}A_{ij}$.

Similarly, if $u_i = (u_{i1}, \cdots, u_{in_i})$ is an n_i-vector, $(i = 1, \cdots, c)$, we define $u = \sum_{i=1}^{c} u_i = u_1 \dotplus \cdots \dotplus u_c$, the *direct sum* of the indicated vectors, to be the $\left(\sum_{i=1}^{c} n_i\right)$-vector $u = (u_{11}, \cdots, u_{1n_1}, u_{21}, \cdots, u_{2n_2}, \cdots, u_{c1}, \cdots, u_{cn_c})$.

In other words, u is the vector obtained by putting the vectors $u_1, \cdots, u_c$ together end to end in the indicated order. If A is the partitioned matrix in 1.7(2), then

$$Au = \sum_{j=1}^{c} A_{1j}u_j \dotplus \sum_{j=1}^{c} A_{2j}u_j \dotplus \cdots \dotplus \sum_{j=1}^{c} A_{dj}u_j$$

$$= \sum_{i=1}^{d} \sum_{j=1}^{c} A_{ij}u_j.$$

1.8 Examples

1.8.1 Let A and B be in $M_n(\mathbf{F})$ and let C be the matrix in $M_{2n}(\mathbf{F})$ defined by

$$C = \begin{pmatrix} A & B \\ B & -A \end{pmatrix}.$$

Then

$$C^2 = \begin{pmatrix} A & B \\ B & -A \end{pmatrix}\begin{pmatrix} A & B \\ B & -A \end{pmatrix} = \begin{pmatrix} A^2 + B^2 & AB - BA \\ BA - AB & A^2 + B^2 \end{pmatrix}.$$

1.8.2 Suppose $A \in M_n(\mathbf{P})$, $(1 \leq k \leq n)$, and $A = \begin{pmatrix} C & D \\ E & K \end{pmatrix}$ where $C \in M_k(\mathbf{P})$, $K \in M_{n-k}(\mathbf{P})$, $E \in M_{n-k,k}(\mathbf{P})$, $D \in M_{k,n-k}(\mathbf{P})$.

Let u be a column vector in $M_{n,1}(\mathbf{P})$ whose first k coordinates are positive and the rest 0 so that $u = v \dotplus 0$ where v has positive coordinates and 0 is an $(n-k)$-tuple of zeros. Then

$$(I_n + A)u = I_n\begin{pmatrix} v \\ 0 \end{pmatrix} + \begin{pmatrix} C & D \\ E & K \end{pmatrix}\begin{pmatrix} v \\ 0 \end{pmatrix} = \begin{pmatrix} v + Cv \\ Ev \end{pmatrix}.$$

Hence $(I_n + A)u$ has at least k positive coordinates; and if it has exactly k positive coordinates, then $Ev = 0$. But if $Ev = 0$ then [see 1.4(2)]

$$\sum_{j=1}^{k} E^{(j)}v_j = 0, \qquad (v_j > 0).$$

Since $E^{(j)}$ has nonnegative coordinates, $(j = 1, \cdots, k)$, it follows that each $E^{(j)}$ is the zero vector and thus that $E = 0_{n-k,k}$.

1.9 Kronecker product

A second important way of putting matrices together is by forming their Kronecker product. If $A \in M_{m,n}(\mathbf{F})$ and $B \in M_{p,q}(\mathbf{F})$, then

$$A \otimes B = \begin{pmatrix} a_{11}B & a_{12}B & \cdots & a_{1n}B \\ a_{21}B & a_{22}B & \cdots & a_{2n}B \\ \cdot & \cdot & & \cdot \\ \cdot & \cdot & & \cdot \\ \cdot & \cdot & & \cdot \\ a_{m1}B & a_{m2}B & \cdots & a_{mn}B \end{pmatrix}$$

The matrix $A \otimes B$ is $mp \times nq$, partitioned into mn blocks and the (i,j) block is the matrix $a_{ij}B \in M_{p,q}(\mathbf{F})$. The matrix $A \otimes B$ is called the *Kronecker product* of A and B and also goes under the name of *direct product* or *tensor product*. Suppose that $C \in M_{n,k}(\mathbf{F})$ and $D \in M_{q,r}(\mathbf{F})$; then $C \otimes D$ is $nq \times kr$ and it follows that the product $P = (A \otimes B)(C \otimes D)$ is an $mp \times kr$ matrix. The block multiplication procedure in **1.7(3)** shows that P can be partitioned into mk $p \times r$ blocks and the (s,t) block is precisely

$$\sum_{j=1}^{n} (a_{sj}B)(c_{jt}D) = \left(\sum_{j=1}^{n} a_{sj}c_{jt} \right) BD.$$

But $\sum_{j=1}^{n} a_{sj}c_{jt}$ is the (s,t) element of the product AC and thus

$$(A \otimes B)(C \otimes D) = AC \otimes BD. \tag{1}$$

In other words, if the products AC and BD exist, then the product $(A \otimes B)(C \otimes D)$ exists and is equal to $AC \otimes BD$.

Several elementary facts about the Kronecker product follow.

1.9.1 $(A \otimes B)^T = A^T \otimes B^T$.

1.9.2 If A and B are over $\mathbf{C}$, then $(\overline{A \otimes B}) = \overline{A} \otimes \overline{B}$, $(A \otimes B)^* = A^* \otimes B^*$.

1.9.3 If A and B are nonsingular, then so is $A \otimes B$ and $(A \otimes B)^{-1} = A^{-1} \otimes B^{-1}$.

1.10 Example

If $X \in M_{m,n}(\mathbf{F})$, $A \in M_{p,m}(\mathbf{F})$, and $B \in M_{n,q}(\mathbf{F})$, then the products $Y = AX$, $Z = XB$, $W = AXB$ are all defined. These ordinary matrix products can be expressed in terms of Kronecker products as follows. With the matrix X associate the mn-vector $x = \sum_{t=1}^{n} \cdot X^{(t)}$. That is, x is the direct sum of the column vectors of X. If y is the pn-vector $y = \sum_{t=1}^{n} \cdot Y^{(t)}$, then, from **1.4(3)**, $y = \sum_{t=1}^{n} \cdot AX^{(t)} = (I_n \otimes A)x$. Similarly, if $z = \sum_{t=1}^{n} \cdot Z^{(t)}$, then, from **1.4(2)**,

$$Z^{(t)} = \sum_{s=1}^{n} b_{st} X^{(s)} = \sum_{s=1}^{n} (B^T)_{ts} I_m X^{(s)}.$$

Hence

$$z = \sum_{t=1}^{q} \cdot \sum_{s=1}^{n} (B^T)_{ts} I_m X^{(s)} = (B^T \otimes I_m)x.$$

If these formulas are put together, then

$$
\begin{aligned}
w = \sum_{t=1}^{q} \cdot W^{(t)} &= (B^T \otimes I_m)(I_n \otimes A)x = (B^T I_n \otimes I_m A)x \\
&= (B^T \otimes A)x.
\end{aligned}
\tag{1}
$$

This formula can be used, for example, to examine solutions to matrix equations such as

$$AX + XA^* = I_n \tag{2}$$

(which arises in stability theory) in which A and X are n-square matrices. For, by means of **1.10(1)**, the equation **1.10(2)** takes the form

$$(I_n \otimes A + \overline{A} \otimes I_n)x = e$$

where $e = \sum_{j=1}^{n} \cdot e_j$ and e_j is the jth column vector of I_n.

2   Numbers Associated with Matrices

2.1 Notation

Before launching into the matrix theory proper it is convenient to have a brief notation available in order to prune the usual foliage of subscripts and superscripts attendant upon this topic.

If $1 \le k \le n$, then $Q_{k,n}$ will denote the totality of strictly increasing sequences of k integers chosen from $1, \cdots, n$. The totality of nondecreasing sequences of k integers chosen from $1, \cdots, n$ will be denoted by $G_{k,n}$. Thus, for example, $Q_{2,3}$ consists of the sequences $(1, 2), (1, 3), (2, 3)$ whereas $G_{2,3}$ consists of the sequences $(1, 1), (1, 2), (1, 3), (2, 2), (2, 3), (3, 3)$. In general, $Q_{k,n}$ has $\binom{n}{k}$ sequences in it and $G_{k,n}$ has $\binom{n+k-1}{k}$ sequences in it. It is useful to have some notation for two other sequence sets. The first is the set $D_{k,n}$ of all sequences of k distinct integers chosen from $1, \cdots, n$. Thus $D_{k,n}$ is obtained from $Q_{k,n}$ by associating with each sequence $\alpha \in Q_{k,n}$ the $k!$ sequences obtained by reordering the integers in all possible ways. Hence $D_{k,n}$ has $\binom{n}{k} k! = n!/(n-k)!$ sequences in it. The other set is $S_{k,n}$ which is simply the totality of sequences of k integers chosen from $1, \cdots, n$. Obviously $S_{k,n}$ has n^k sequences in it. If $a_1, \cdots, a_n$ are n elements in $\mathbf{F}$ and $\omega = (i_1, \cdots, i_k)$ is a sequence in either of $Q_{k,n}$ or $G_{k,n}$, then it is sometimes convenient to designate the product $a_{i_1} a_{i_2} \cdots a_{i_k}$ by a_ω. In this notation it is easy to write down the definition of the *elementary symmetric function* of degree k of the n elements $a_1, \cdots, a_n$: it is

$$E_k(a) = E_k(a_1, a_2, \cdots, a_n) = \sum_{\omega \in Q_{k,n}} a_\omega. \tag{1}$$

For example, $E_1(a_1, a_2, a_3) = a_1 + a_2 + a_3$, $E_2(a_1, a_2, a_3) = a_1 a_2 + a_1 a_3 + a_2 a_3$, and $E_3(a_1, a_2, a_3) = a_1 a_2 a_3$. It is useful to assign an ordering to the sequences in $S_{k,n}$ (and therefore in $Q_{k,n}, G_{k,n}, D_{k,n}$). The most convenient one is the *lexicographic ordering* defined as follows: if α and β are sequences in $S_{k,n}$, then α *precedes* β, written $\alpha < \beta$, if there exists an integer t, $(1 \leq t \leq k)$, for which $\alpha_1 = \beta_1, \cdots, \alpha_{t-1} = \beta_{t-1}, \alpha_t < \beta_t$. Thus $(3, 5, 8)$ precedes $(4, 5, 6)$ in $Q_{3,8}$ and $(3, 3, 4)$ precedes $(3, 4, 4)$ in $G_{3,4}$.

2.2 Submatrices

Suppose $A = (a_{ij}) \in M_{m,n}(\mathbf{F})$; k and r are positive integers satisfying $1 \leq k \leq m, 1 \leq r \leq n$; and $\alpha = (i_1, \cdots, i_k) \in Q_{k,m}, \beta = (j_1, \cdots, j_r) \in Q_{r,n}$. Then the matrix $B \in M_{k,r}(\mathbf{F})$ whose (s,t) entry is $a_{i_s j_t}$, $(s = 1, \cdots, k;$ $t = 1, \cdots, r)$, is called the *submatrix of A lying in rows α and columns β* and we designate B with the notation

$$B = A[\alpha|\beta]. \tag{1}$$

For example, if $A = (a_{ij}) \in M_{5,6}(\mathbf{F})$, $k = 2, r = 3, \alpha = (1, 3), \beta = (2, 3, 4)$, then

$$A[\alpha|\beta] = \begin{pmatrix} a_{12} & a_{13} & a_{14} \\ a_{32} & a_{33} & a_{34} \end{pmatrix} \in M_{2,3}(\mathbf{F}). \tag{2}$$

We use the notation $A(\alpha|\beta]$ to designate the submatrix of A whose rows are precisely those complementary to α and whose columns are designated by β. Similarly, $A[\alpha|\beta]$ includes rows α and excludes columns β, whereas $A(\alpha|\beta)$ excludes rows α and columns β. Thus $A(\alpha|\beta] \in M_{m-k,r}(\mathbf{F})$, $A[\alpha|\beta) \in M_{k,n-r}(\mathbf{F})$, and $A(\alpha|\beta) \in M_{m-k,n-r}(\mathbf{F})$.

In the example used in 2.2(2)

$$A(\alpha|\beta] = \begin{pmatrix} a_{22} & a_{23} & a_{24} \\ a_{42} & a_{43} & a_{44} \\ a_{52} & a_{53} & a_{54} \end{pmatrix} \in M_{3,3}(\mathbf{F});$$

$$A[\alpha|\beta) = \begin{pmatrix} a_{11} & a_{15} & a_{16} \\ a_{31} & a_{35} & a_{36} \end{pmatrix} \in M_{2,3}(\mathbf{F});$$

$$A(\alpha|\beta) = \begin{pmatrix} a_{21} & a_{25} & a_{26} \\ a_{41} & a_{45} & a_{46} \\ a_{51} & a_{55} & a_{56} \end{pmatrix} \in M_{3,3}(\mathbf{F}).$$

We can also extend the definition of $A[\alpha|\beta]$ to the case in which α and β are sequences in $D_{k,n}$, $S_{k,n}$ or $G_{k,n}$ by simply allowing other than the natural order for row and column choices and permitting repetitions in these choices. For example

$$A[2, 1|4, 4] = \begin{pmatrix} a_{24} & a_{24} \\ a_{14} & a_{14} \end{pmatrix}.$$

If $A \in M_n(\mathbf{F})$, then $A[\alpha|\alpha]$, $\alpha \in Q_{r,n}$, will be called an r-square *principal submatrix* of A.

2.3 Permutations

The notation S_n will be used to designate the *symmetric group* of degree n on the n integers $1, \cdots, n$. That is, S_n is the totality of one-one functions, or *permutations*, on the set $\{1, \cdots, n\}$ onto itself. The set S_n has $n!$ elements in it. A *cycle* in S_n is a permutation σ which has the property that there exists a subset of $\{1, \cdots, n\}$, call it $\{i_1, \cdots, i_k\}$, such that $\sigma(i_1) = i_2$, $\sigma(i_2) = i_3, \cdots, \sigma(i_{k-1}) = i_k, \sigma(i_k) = i_1$, and $\sigma(j) = j$ for $j \neq i_t$, $(t = 1, \cdots, k)$. The integer k is called the *length* of the cycle and the cycles of length 2 are called *transpositions*. The important fact about transpositions is that any $\sigma \in S_n$ is the product (in the ordinary function composition sense) of transpositions. It is very simple to see that any permutation σ is a product of cycles acting on disjoint subsets of $\{1, \cdots, n\}$ (disjoint cycles) and this factorization is unique to within order. If the lengths of these cycles are $p_1, \cdots, p_m$, then σ is said to have the *cycle structure* $[p_1, \cdots, p_m]$ (some of the p_i may be 1). Although the factorization into a product of transpositions is not unique, it is true that any two such factorizations of the same permutation σ must both have an even or both have an odd number of transpositions: σ is called *even* or *odd* accordingly. The *sign* of σ is defined by

$$\text{sgn } \sigma = \begin{cases} 1 \text{ if } \sigma \text{ is even} \\ -1 \text{ if } \sigma \text{ is odd.} \end{cases}$$

If $A \in M_n(\mathbf{F})$ and $\sigma \in S_n$, then the sequence of elements $a_{1\sigma(1)}, \cdots, a_{n\sigma(n)}$ will be called the *diagonal* of A corresponding to σ. If σ is the identity permutation, i.e., $\sigma(j) = j$, $(j = 1, \cdots, n)$, then the diagonal corresponding to σ, namely $a_{11}, \cdots, a_{nn}$, is called the *main diagonal* of A. A matrix $A \in M_n(\mathbf{F})$ for which $a_{i\sigma(i)} = 1$, $(i = 1, \cdots, n)$, and $a_{ij} = 0$ otherwise, is called a *permutation matrix*.

2.4 Determinants

If $A = (a_{ij}) \in M_n(\mathbf{F})$, then the *determinant* of A is defined by

$$d(A) = \sum_{\sigma \in S_n} \text{sgn } \sigma \prod_{i=1}^{n} a_{\sigma(i)i}. \tag{1}$$

Thus the determinant is the sum of the products of the elements in all $n!$ diagonals each weighted with ± 1 according as the diagonal corresponds to an even or an odd permutation $\sigma \in S_n$. Other notations used for $d(A)$ are $\det (A)$, $D(A)$, $|A|$. The determinant of a principal submatrix of A is called a *principal subdeterminant*. In general, the determinant of any square submatrix is called a *subdeterminant*.

The following list contains most of the important elementary facts about determinants.

In what follows, A and B are matrices in $M_n(\mathbf{F})$.

2.4.1 $d(A^T) = d(A)$, $d(A^*) = \overline{d(A)}$, $d(cA) = c^n d(A)$ for any scalar c.

2.4.2 If B is the matrix obtained from A by permuting the rows (columns) of A according to a permutation $\sigma \in S_n$, i.e., $B_{(i)} = A_{(\sigma(i))}$, $(B^{(i)} = A^{(\sigma(i))})$, $(i = 1, \cdots, n)$, then $d(B) = \text{sgn } \sigma \, d(A)$.

2.4.3 If B is obtained from A by multiplying a fixed row (column) of A by a scalar c, then $d(B) = cd(A)$.

2.4.4 If there exist scalars $c_1, \cdots, c_n$, not all 0, such that $\sum_{j=1}^{n} c_j A_{(j)} = 0_n$ or $\sum_{j=1}^{n} c_j A^{(j)} = 0_n$, then $d(A) = 0$. It follows from this that if any row or column of A consists only of zeros, then $d(A) = 0$; moreover, if A has two identical rows or two identical columns, then $d(A) = 0$.

2.4.5 $d(AB) = d(A)d(B)$. If A^{-1} exists, then $d(A) \neq 0$ and $d(A^{-1}) = (d(A))^{-1}$; conversely, if $d(A) \neq 0$, then A is nonsingular; $d(I_n) = 1$; $d(0_{n,n}) = 0$.

2.4.6 If for some v, $1 \leq p \leq n$, $B_{(i)} = A_{(i)}$, $(i \neq p)$, and

$$B_{(p)} = A_{(p)} + \sum_{j=1, j \neq p}^{n} c_j A_{(j)}, \qquad (c_j \in \mathbf{F}),$$

then $d(B) = d(A)$. Similarly, if $B^{(i)} = A^{(i)}$, $(i \neq p)$, and

$$B^{(p)} = A^{(p)} + \sum_{j=1, j \neq p}^{n} c_j A^{(j)},$$

then $d(B) = d(A)$. In other words, adding multiples of other rows (columns) to a fixed row (column) does not alter the determinant of a matrix.

2.4.7 If $A_{(p)} = \sum_{j=1}^{k} c_j z_j$ where $z_j \in M_{1,n}(\mathbf{F})$ and $c_j \in \mathbf{F}$, $(j = 1, \cdots, k)$, and $Z_j \in M_n(\mathbf{F})$ is the matrix whose pth row is z_j and whose rth row is $A_{(r)}$, $(r = 1, \cdots, n; r \neq p)$, then $d(A) = \sum_{j=1}^{k} c_j d(Z_j)$. This latter formula together with 2.4.2 says that the function $d(A)$ is an *alternating multilinear* function of the rows.

2.4.8 (Expansion by a row or column)

$$d(A) = \sum_{j=1}^{n} (-1)^{i+j} a_{ij} d(A(i|j))$$

$$= \sum_{i=1}^{n} (-1)^{i+j} a_{ij} d(A(i|j)).$$

2.4.9 If $A \in M_n(\mathbf{F})$ and $B = (b_{ij})$ where $b_{ij} = (-1)^{i+j} d(A(j|i))$, $(i, j = 1, \cdots, n)$, then $BA = AB = d(A) I_n$. The matrix B is called the *adjoint* of A, written adj (A). Some authors call B the *adjugate* of A.

2.4.10 Let v, k, λ be positive integers satisfying $0 < \lambda < k < v$. A matrix $A \in M_v(\mathbf{I})$ with zeros and ones as entries satisfying $AA^T = (k - \lambda)I_v + \lambda J$ where $J \in M_v(\mathbf{I})$ has every entry 1 is called a *(v,k,λ)-matrix*. The entries of A are 0 and 1 and hence $(AA^T)_{ii} = \sum_{t=1}^{v} a_{it}^2$ is just the sum of the entries in row i. Hence $AJ = kJ$,

$$AA^T = (k - \lambda)I_v + \frac{\lambda}{k} AJ,$$

$$A\left(A^T - \frac{\lambda}{k} J\right) = (k - \lambda)I_v,$$

$$A^{-1} = \frac{1}{k - \lambda}\left(A^T - \frac{\lambda}{k} J\right) = \frac{1}{k(k - \lambda)} (kA^T - \lambda J).$$

It follows that the adjoint of A is the matrix

$$B = \frac{d(A)}{k(k - \lambda)} (kA^T - \lambda J).$$

We shall also derive an interesting relationship connecting the integers v, k, and λ. We have

$A^T A = A^{-1}(AA^T)A = A^{-1}((k - \lambda)I_v + \lambda J)A = (k - \lambda)I_v + \lambda A^{-1}JA.$

From $AJ = kJ$ we have $A^T A = (k - \lambda)I_v + \lambda k^{-1}JA.$ Thus $JA^T A = (k - \lambda)J + \lambda k^{-1}vJA$ (note that $J^2 = vJ$). Now $(AJ)^T = kJ$ so $kJA = (k - \lambda)J + \lambda k^{-1}vJA.$ Thus $JA = cJ$ where $c = (k - \lambda)/(k - \lambda k^{-1}v).$ Then $JAJ = cJ^2 = cvJ,$ $kJ^2 = cvJ,$ $k = c$ and finally we have $\lambda = k(k - 1)/(v - 1).$

2.4.11 (Laplace expansion theorem) Let $1 \le r \le n,$ let α be a fixed sequence in $Q_{r,n},$ and for any $\beta = (\beta_1, \cdots, \beta_r) \in Q_{r,n}$ let $s(\beta) = \sum\limits_{t=1}^{r} \beta_t.$ Then

$$d(A) = (-1)^{s(\alpha)} \sum_{\beta \in Q_{r,n}} (-1)^{s(\beta)} d(A[\alpha|\beta]) d(A(\alpha|\beta)).$$

2.4.12 If $A \in M_p(\mathbf{F}),$ $B \in M_q(\mathbf{F}),$ $C \in M_{p,q}(\mathbf{F}),$ and $H = \begin{pmatrix} A & C \\ 0_{q,p} & B \end{pmatrix},$ then $d(H) = d(A)d(B).$ It follows that if

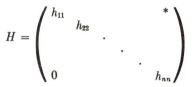

where the * indicates that the elements above the main diagonal are unspecified and the 0 indicates that all elements below the main diagonal are zero, then $d(H) = \prod\limits_{i=1}^{n} h_{ii}.$

2.4.13 If $A \in M_p(\mathbf{F})$ and $B \in M_q(\mathbf{F}),$ then $A \otimes B \in M_{pq}(\mathbf{F})$ and $d(A \otimes B) = d^q(A)d^p(B).$

2.4.14 (Binet-Cauchy theorem) Suppose $A \in M_{n,p}(\mathbf{F}),$ $B \in M_{p,m}(\mathbf{F}),$ and $C = AB \in M_{n,m}(\mathbf{F}).$ If $1 \le r \le \min(n, m, p),$ $\alpha \in Q_{r,n},$ $\beta \in Q_{r,m},$ then $d(C[\alpha|\beta]) = \sum\limits_{\omega \in Q_{r,p}} d(A[\alpha|\omega]) d(B[\omega|\beta]).$

2.5   The quadratic relations among subdeterminants

If $A \in M_{k,n}(\mathbf{F}),$ where $\mathbf{F}$ is here assumed to be either $\mathbf{R}$ or $\mathbf{C},$ then the subdeterminants $d(A[1, \cdots, k|\omega]),$ $\omega \in S_{k,n}$ are related to one another. There are certain quadratic relations that must obtain among them. Suppose $p(x_1, \cdots, x_k)$ is a function of k independent variables $x_1, \cdots, x_k$ defined for positive integer values of the variables and taking on values in $\mathbf{F}.$ Suppose moreover that

(i) $p(x_{\sigma(1)}, \cdots, x_{\sigma(k)}) = \operatorname{sgn} \sigma\, p(x_1, \cdots, x_k),$ $\sigma \in S_k,$ (p is alternating);

(ii) $p(x_1, \cdots, x_k)$ is not 0 for all choices of positive integers $x_1, \cdots, x_k$ (p is not identically 0);

(iii) p satisfies

$$p(\alpha_1, \cdots, \alpha_k)p(\beta_1, \cdots, \beta_k)$$

$$= \sum_{t=1}^{k} p(\alpha_1, \cdots, \alpha_{s-1}, \beta_t, \alpha_{s+1}, \cdots, \alpha_k)p(\beta_1, \cdots, \beta_{t-1}, \alpha_s, \beta_{t+1}, \cdots, \beta_k)$$

for each $s = 1, \cdots, k$, and any sequences $(\alpha_1, \cdots, \alpha_k)$, $(\beta_1, \cdots, \beta_k)$ in $D_{k,n}$. Then there exists a matrix $A \in M_{k,n}(\mathbf{F})$ such that

$$d(A[1, \cdots, k|i_1, \cdots, i_k]) = p(i_1, \cdots, i_k)$$

for every sequence $(i_1, \cdots, i_k) \in Q_{k,n}$. Conversely, given a matrix $A \in M_{k,n}(\mathbf{F})$, if one defines a function $p(i_1, \cdots, i_k)$ by

$$p(i_1, \cdots, i_k) = d(A[1, \cdots, k|i_1, \cdots, i_k])$$

for arbitrary sequences $(i_1, \cdots, i_k) \in S_{k,n}$, then p satisfies the quadratic relations **2.5(iii)** and p alternates in the sense that $p(i_{\sigma(1)}, \cdots, i_{\sigma(k)}) = \text{sgn } \sigma \, p(i_1, \cdots, i_k)$ for any $\sigma \in S_k$.

2.6 Examples

2.6.1 If $A \in M_{2,n}(\mathbf{F})$, then the choices of $\alpha: 1 \leq \alpha_1 < \alpha_2 \leq n$, and $\beta: 1 \leq \beta_1 < \beta_2 \leq n$ are the only ones for which the relations **2.5(iii)** do not collapse to something trivial. In this case these quadratic relations become

$$p(\alpha_1, \alpha_2)p(\beta_1, \beta_2) = p(\beta_1, \alpha_2)p(\alpha_1, \beta_2) + p(\beta_2, \alpha_2)p(\beta_1, \alpha_1), \qquad (1)$$

where

$$p(i_1, i_2) = d(A[1, 2|i_1, i_2]).$$

If there is any overlap in the sequences α and β, then **2.6(1)** does not give us any information. For example, if $\alpha_1 = \beta_1$, then the left side becomes $p(\alpha_1, \alpha_2)p(\alpha_1, \beta_2)$, whereas the right side becomes $p(\alpha_1, \alpha_2)p(\alpha_1, \beta_2) + p(\beta_2, \alpha_2)p(\beta_1, \alpha_1)$; and the alternating property implies that $p(\alpha_1, \alpha_1) = 0$. It follows from **2.6(1)** that not all of the 2-square subdeterminants of a 2×4 matrix can have a common nonzero value c, otherwise we would have $c^2 = 2c^2$.

2.6.2 (**Vandermonde matrix**) Let $r_1, \cdots, r_n$ be elements of $\mathbf{F}$ and suppose $A = (a_{ij}) = (r_j^{i-1}) \in M_n(\mathbf{F})$. A is called the *Vandermonde* or *alternant* matrix. Let B be the matrix defined by

$$B_{(1)} = A_{(1)}, \; B_{(i)} = A_{(i)} - r_1 A_{(i-1)}, \qquad (i = 2, \ldots, n).$$

Then

$$B = \begin{pmatrix} 1 & 1 & \cdots & 1 \\ 0 & r_2 - r_1 & \cdots & r_n - r_1 \\ 0 & r_2^2 - r_1 r_2 & \cdots & r_n^2 - r_1 r_n \\ \cdot & \cdot & & \cdot \\ \cdot & \cdot & & \cdot \\ \cdot & \cdot & & \cdot \\ 0 & r_2^{n-1} - r_1 r_2^{n-2} & \cdots & r_n^{n-1} - r_1 r_n^{n-2} \end{pmatrix}$$

and according to 2.4.6, 2.4.8, and 2.4.3 in succession it follows that

$$d(A) = d(B) = d(B(1|1))$$

$$= (r_2 - r_1) \cdots (r_n - r_1)d \begin{pmatrix} 1 & \cdots & 1 \\ r_2 & \cdots & r_n \\ r_2^2 & \cdots & r_n^2 \\ . & & . \\ . & & . \\ . & & . \\ r_2^{n-2} & \cdots & r_n^{n-2} \end{pmatrix}$$

This last matrix on the right has exactly the same structure as A only it is $(n-1)$-square and involves $r_2, \cdots, r_n$. The argument can be repeated to obtain finally

$$d(A) = \prod_{n \geq j > i \geq 1} (r_j - r_i).$$

2.6.3 If $A \in M_n(\mathbf{F})$ contains an $s \times t$ submatrix consisting of zero entries where $s + t = n + 1$, then $d(A) = 0$. For, let B be the matrix obtained from A by permuting the rows and columns of A in such a way as to bring the submatrix of zero entries into the upper left corner:

$$B = \begin{pmatrix} 0_{s,t} & C \\ D & K \end{pmatrix}, \qquad (C \in M_{s,n-t}(\mathbf{F})).$$

If $b_{1\sigma(1)}, \cdots, b_{n\sigma(n)}$ is a diagonal of B, then $\prod_{i=1}^{n} b_{i\sigma(i)} = 0$ unless the s distinct integers $\sigma(1), \cdots, \sigma(s)$ are among the $n - t$ integers $t + 1, \cdots, n$. In order for this to be possible, $n - t \geq s$ or $n \geq s + t = n + 1$. Thus every product $\prod_{i=1}^{n} b_{i\sigma(i)}$ is zero and $d(B) = 0$. But $d(A) = \pm d(B)$, by **2.4.2**, and hence $d(A) = 0$.

2.7 Compound matrices

If $A \in M_{m,n}(\mathbf{F})$ and $1 \leq r \leq \min{(m,n)}$, then the rth *compound matrix* or rth *adjugate* of A is the $\binom{m}{r} \times \binom{n}{r}$ matrix whose entries are $d(A[\alpha|\beta])$, $\alpha \in Q_{r,m}, \beta \in Q_{r,n}$ arranged lexicographically in α and β. This matrix will be designated by $C_r(A)$. Another notation for it is $|A|^{(r)}$. For example, if $A = (a_{ij}) \in M_3(\mathbf{F})$ and $r = 2$, then

$$C_2(A) = \begin{pmatrix} d(A[1,2|1,2]) & d(A[1,2|1,3]) & d(A[1,2|2,3]) \\ d(A[1,3|1,2]) & d(A[1,3|1,3]) & d(A[1,3|2,3]) \\ d(A[2,3|1,2]) & d(A[2,3|1,3]) & d(A[2,3|2,3]) \end{pmatrix}.$$

2.7.1 If $A \in M_{m,n}(\mathbf{F})$ and $B \in M_{n,k}(\mathbf{F})$ and $1 \leq r \leq \min{(m,n,k)}$, then

$$C_r(AB) = C_r(A)C_r(B). \tag{1}$$

The formula 2.7(1) is just another way of writing down the Binet-Cauchy theorem 2.4.14.

2.7.2 Some important facts about $C_r(A)$, $A \in M_n(\mathbf{F})$, $1 \leq r \leq n$, follow. If A is nonsingular, then,

$$(C_r(A))^{-1} = C_r(A^{-1}). \tag{2}$$

$$\begin{aligned} C_r(A^*) &= (C_r(A))^*, \\ C_r(A^T) &= (C_r(A))^T, \\ C_r(\overline{A}) &= \overline{C_r(A)}. \end{aligned} \tag{3}$$

For any $k \in \mathbf{F}$

$$C_r(kA) = k^r C_r(A), \tag{4}$$

$$C_r(I_n) = I_{\binom{n}{r}}. \tag{5}$$

(Sylvester-Franke theorem)

$$d(C_r(A)) = (d(A))^{\binom{n-1}{r-1}} \tag{6}$$

2.7.3 If $A \in M_{r,n}(\mathbf{F})$, and the r rows of A are denoted by $u_1, \cdots, u_r$ in succession, $(1 \leq r \leq n)$, then $C_r(A)$ is an $\binom{n}{r}$-tuple and is sometimes called the *Grassmann product* or *skew-symmetric product* of the vectors $u_1, \cdots, u_r$. The usual notation for this $\binom{n}{r}$-tuple of subdeterminants of A is $u_1 \wedge \cdots \wedge u_r$. By the properties of determinants, if $\sigma \in S_r$, then

$$u_{\sigma(1)} \wedge \cdots \wedge u_{\sigma(r)} = \operatorname{sgn} \sigma\, u_1 \wedge \cdots \wedge u_r. \tag{7}$$

If $B \in M_n(\mathbf{F})$, then

$$C_r(B)u_1 \wedge \cdots \wedge u_r = Bu_1 \wedge \cdots \wedge Bu_r. \tag{8}$$

2.8 Symmetric functions; trace

If $A = (a_{ij}) \in M_n(\mathbf{F})$ and $1 \leq r \leq n$, then we designate the *sum of all the* $\binom{n}{r}$ *r-square principal subdeterminants* by

$$E_r(A) = \sum_{\omega \in Q_{r,n}} d(A[\omega|\omega]). \tag{1}$$

Thus $E_n(A) = d(A)$, for example. The function $E_1(A) = \sum_{i=1}^n a_{ii}$ is called the *trace* of A and a special notation is used for it:

$$E_1(A) = \operatorname{tr}(A) = \sum_{i=1}^n a_{ii}. \tag{2}$$

Some properties of the trace follow.

2.8.1   **(Linearity)** If A and B are in $M_n(\mathbf{F})$ and c and d are in $\mathbf{F}$, then

$$\operatorname{tr}(cA + dB) = c\operatorname{tr}(A) + d\operatorname{tr}(B).$$

2.8.2   If $A = (a_{ij}) \in M_{m,n}(\mathbf{C})$, then

$$\operatorname{tr}(AA^*) = \operatorname{tr}(A^*A) = \sum_{i=1}^{m}\sum_{j=1}^{n}|a_{ij}|^2. \tag{3}$$

The number $(\operatorname{tr}(AA^*))^{1/2}$ is called the *Euclidean norm* of A.

2.8.3   If A and B are in $M_n(\mathbf{F})$, then $\operatorname{tr}(AB) = \operatorname{tr}(BA)$. More generally, if $A_p \in M_n(\mathbf{F})$, $(p = 1, \cdots, k)$, and $\sigma \in S_k$ has the form $\sigma(t) = t + r$ (mod k), for some fixed r, $(1 \le r \le k)$, then

$$\operatorname{tr}\left(\prod_{p=1}^{k} A_p\right) = \operatorname{tr}\left(\prod_{p=1}^{k} A_{\sigma(p)}\right).$$

2.8.4   If $A \in M_n(\mathbf{F})$ and $S \in M_n(\mathbf{F})$ and S is nonsingular, then

$$\operatorname{tr}(S^{-1}AS) = \operatorname{tr}(A).$$

2.8.5 $\operatorname{tr}(A \otimes B) = \operatorname{tr}(A)\operatorname{tr}(B).$

2.8.6 $\operatorname{tr}(C_r(A)) = E_r(A).$

2.9 Permanents

Another scalar-valued function of a matrix $A \in M_n(\mathbf{F})$ is the *permanent* of A which will be designated here by the notation $p(A)$:

$$p(A) = \sum_{\sigma \in S_n} \prod_{i=1}^{n} a_{\sigma(i)i}. \tag{1}$$

The permanent is also known as the "plus determinant" and is sometimes denoted by $\overset{+}{|}A\overset{+}{|}$.

2.9.1   **(Incidence matrices)** Suppose $P = \{p_1, \cdots, p_n\}$ is a set of n items and suppose that $P_1, \cdots, P_n$ are n subsets of P. This situation is described completely by an incidence matrix $A \in M_n(\mathbf{I})$ whose (i,j) entry is 1 or 0 according as $p_i \in P_j$ or not. An ordering of the elements of P, $p_{\sigma(1)}, \cdots, p_{\sigma(n)}$ with $p_{\sigma(i)} \in P_i$, $(i = 1, \cdots, n)$, is called a *system of distinct representatives*. For example, the set $P = \{p_1, \cdots, p_n\}$ could be a group with the sets P_i constituting a coset decomposition of P. Then a system of distinct representatives is just an ordered selection of the p_i, one from each coset. A term $\prod_{i=1}^{n} a_{\sigma(i)i}$ of $p(A)$ is 1, if and only if $p_{\sigma(i)} \in P_i$, $(i = 1, \cdots, n)$. Thus $p(A)$ is just a count of the total number of systems of distinct representatives.

2.10 Example

Suppose the distinct points c_i, $(i = 1, \cdots, n)$, are connected to one another. At time zero, n particles q_i are located one at each point c_i, $(i = 1, \cdots, n)$. The following information concerning the motion of the particles is known: the probability that particle q_i moves from point c_i to point c_j is p_{ij}, $(i = 1, \cdots, n; j = 1, \cdots, n)$. Let $P = (p_{ij}) \in M_n(\mathbf{C})$. Question: what is the probability that the ultimate arrangement of the particles is such that there is precisely one particle at each of the points c_i, $(i = 1, \cdots, n)$? The answer is $p(P)$. For, $\prod_{i=1}^{n} p_{\sigma(i),i}$ is the probability that particles $q_{\sigma(1)}, \cdots, q_{\sigma(n)}$ move to points $c_1, \cdots, c_n$ in that order, and $p(P) = \sum_{\sigma \in S_n} \prod_{i=1}^{n} p_{\sigma(i),i}$ is thus just the probability that the particles distribute themselves one at each point in *some* order.

2.11 Properties of permanents

Some properties of $p(A)$ for $A \in M_n(\mathbf{F})$ follow.

2.11.1 If D and L are diagonal matrices and P and Q are permutation matrices (all in $M_n(\mathbf{F})$), then $p(PAQ) = p(A)$, $p(DAL) = \left(\prod_{i=1}^{n} D_{ii}L_{ii} \right) p(A)$.

2.11.2 $p(A^T) = p(A)$, $p(A^*) = \overline{p(A)}$, $p(cA) = c^n p(A)$ for any scalar c.

2.11.3 $p(P) = 1$ for any permutation matrix $P \in M_n(\mathbf{F})$, $p(0_{n,n}) = 0$.

2.11.4 If for some p, $A_{(p)} = \sum_{j=1}^{k} c_j z_j$, where $z_j \in M_{1,n}(\mathbf{F})$, $c_j \in \mathbf{F}$, $(j = 1, \cdots, n)$, and $Z_j \in M_n(\mathbf{F})$ is the matrix whose pth row is z_j and whose rth row is $A_{(r)}$, $(r \neq p)$, then

$$p(A) = \sum_{j=1}^{n} c_j p(Z_j).$$

In other words, $p(A)$ is a multilinear function of the rows (and columns, in view of **2.11.2**) of A.

2.11.5 (Expansion by a row or column)

$$p(A) = \sum_{j=1}^{n} a_{ij} p(A(i|j))$$
$$= \sum_{i=1}^{n} a_{ij} p(A(i|j)).$$

2.11.6 (Laplace expansion theorem for permanents) Let $1 \leq r \leq n$ and let α be a fixed sequence in $Q_{r,n}$. Then

$$p(A) = \sum_{\beta \in Q_{r,n}} p(A[\alpha|\beta])p(A(\alpha|\beta)).$$

2.11.7 (Binet-Cauchy theorem for permanents) Let $A \in M_{m,n}(\mathbf{F})$, $B \in M_{n,m}(\mathbf{F})$, and $C = AB \in M_m(\mathbf{F})$, $(1 \leq m \leq n)$, then

$$p(C) = \sum_{\alpha \in G_{m,n}} \frac{p(A[1, \cdots, m|\alpha])p(B[\alpha|1, \cdots, m])}{\mu(\alpha)}.$$

The number $\mu(\alpha)$ is defined to be the product of the factorials of the multiplicities of the distinct integers appearing in the sequence α; e.g., $\mu(1, 1, 3, 3, 3, 5) = 2!3!1!$

2.12 Induced matrices

If $A \in M_{m,n}(\mathbf{C})$ and $1 \leq r \leq n$, then the rth *induced matrix* of A, designated by $P_r(A)$ is the $\binom{m+r-1}{r} \times \binom{n+r-1}{r}$ matrix whose entries are $p(A[\alpha|\beta])/\sqrt{\mu(\alpha)\mu(\beta)}$ for $\alpha \in G_{r,m}$ and $\beta \in G_{r,n}$, arranged lexicographically in α and β. For example, if $n = 3$, $A = (a_{ij}) \in M_2(\mathbf{F})$, and $r = 2$, then $\binom{n+r-1}{r} = 3$ and

$$P_2(A) = \begin{pmatrix} \dfrac{p(A[1,1|1,1])}{2} & \dfrac{p(A[1,1|1,2])}{\sqrt{2}} & \dfrac{p(A[1,1|2,2])}{2} \\[2mm] \dfrac{p(A[1,2|1,1])}{\sqrt{2}} & p(A[1,2|1,2]) & \dfrac{p(A[1,2|2,2])}{\sqrt{2}} \\[2mm] \dfrac{p(A[2,2|1,1])}{2} & \dfrac{p(A[2,2|1,2])}{\sqrt{2}} & \dfrac{p(A[2,2|2,2])}{2} \end{pmatrix}.$$

Some properties of the induced matrix for $m = n$ follow.

2.12.1 If $A \in M_n(\mathbf{F})$, then $p(A)$ always appears as the $(1, \cdots, n; 1, \cdots, n)$ entry in $P_n(A)$.

2.12.2 $P_r(AB) = P_r(A)P_r(B)$ for A and B in $M_n(\mathbf{F})$.

2.12.3 If $A \in M_n(\mathbf{C})$, then $P_r(A^*) = (P_r(A))^*$, $P_r(A^T) = (P_r(A))^T$, $P_r(\overline{A}) = \overline{P_r(A)}$, $P_r(A^{-1}) = (P_r(A))^{-1}$ when A is nonsingular.

2.12.4 $P_r(kA) = k^r P_r(A)$ for any $k \in \mathbf{F}$.

2.12.5 $P_r(I_n) = I_{\binom{n+r-1}{r}}$.

2.12.6 $d(P_r(A)) = (d(A))^{\binom{n+r-1}{r-1}}$.

2.12.7   If $u_1, \cdots, u_r$ are n-vectors, then the $\binom{n+r-1}{r}$-tuple whose α coordinate is

$$\frac{p(A[1, \cdots, r|\alpha])}{\sqrt{\mu(\alpha)}}.$$

arranged lexicographically in $\alpha \in G_{r,n}$ is called the *symmetric product* of the vectors $u_1, \cdots, u_r$ and is denoted by $u_1 \cdots u_r$.

If $\sigma \in S_r$, then $u_{\sigma(1)} \cdots u_{\sigma(r)} = u_1 \cdots u_r$.

If $A \in M_n(\mathbf{F})$, then

$$P_r(A)u_1 \cdots u_r = (Au_1) \cdots (Au_r). \tag{1}$$

2.13 Characteristic polynomial

2.13.1   Let $\mathbf{F}$ be the field $\mathbf{R}$ or $\mathbf{C}$. Let $\lambda_1, \cdots, \lambda_k$ be indeterminates over $\mathbf{F}$ and let

$$p(\lambda_1, \cdots, \lambda_k) = \sum_{i_1 + \cdots + i_k \leq m} c_{i_1 \cdots i_k} \lambda_1^{i_1} \cdots \lambda_k^{i_k}, \qquad (c_{i_1 \cdots i_k} \in \mathbf{F}),$$

be a polynomial in $\lambda_1, \cdots, \lambda_k$ with coefficients in $\mathbf{F}$. Let $A_i \in M_n(\mathbf{F})$, $(i = 1, \cdots, k)$. Then $p(A_1, \cdots, A_k)$ is the matrix

$$\sum_{i_1 + \cdots + i_k \leq m} c_{i_1 \cdots i_k} A_1^{i_1} \cdots A_k^{i_k},$$

(A_i^0 is taken to be I_n). In particular, if $k = 1$, $\lambda_1 = \lambda$, $A \in M_n(\mathbf{F})$, and $p(\lambda) = \sum_{i=0}^{m} a_i \lambda^i$, then $p(A) = \sum_{i=0}^{m} a_i A^i$.

2.13.2   If A is an n-square matrix over $\mathbf{C}$ and λ is an indeterminate over $\mathbf{C}$, then the polynomial $p(\lambda) = d(\lambda I_n - A)$ is called the *characteristic polynomial* of A. The matrix $(\lambda I_n - A) \in M_n(\mathbf{C}[\lambda])$ is called the *characteristic matrix* of A. In terms of the numbers $E_r(A)$ (see **2.8**) the characteristic polynomial can be written

$$p(\lambda) = d(\lambda I_n - A) = \sum_{t=0}^{n} \lambda^{n-t} (-1)^t E_t(A). \tag{1}$$

2.14 Examples

2.14.1   The arguments necessary to establish the formula **2.13(1)** constitute an interesting example of the use of some of the items we have introduced. If a term is to involve λ^{n-t} in $d(\lambda I_n - A)$, it must arise in the product of the terms of a diagonal δ corresponding to a permutation σ having $n - t$ elements along the main diagonal; say $\lambda - a_{jj}$, $(j = i_1, \cdots, i_{n-t};\ 1 \leq i_1 < i_2 < \cdots < i_{n-t} \leq n)$. Then the product

$$\prod_{j=i_1}^{i_{n-t}} (\lambda - a_{jj}) \tag{1}$$

yields the term λ^{n-t}. The rest of δ can be completed by using a diagonal in the rows and columns not already used; that is, using a diagonal from

$$(\lambda I_n - A)(i_1, \cdots, i_{n-t} | i_1, \cdots, i_{n-t}),$$

corresponding to a permutation φ of the set complementary to $i_1, \cdots, i_{n-t}$. But since σ holds $i_1, \cdots, i_{n-t}$ fixed, it is clear that sgn σ = sgn φ. Moreover, we want the coefficient of λ^{n-t}, so we can only choose constant terms (not involving λ) in

$$(\lambda I_n - A)(i_1, \cdots, i_{n-t} | i_1, \cdots, i_{n-t})$$

to complete δ. Thus the coefficient of λ^{n-t} that arises from the product **2.14(1)** is just

$$d(-A(i_1, \cdots, i_{n-t} | i_1, \cdots, i_{n-t})) = (-1)^t d(A(i_1, \cdots, i_{n-t} | i_1, \cdots, i_{n-t})).$$

Each choice of integers $\beta \in Q_{n-t,n}$ thereby yields a term $\lambda^{n-t}(-1)^t d(A[\alpha|\alpha])$ where $\alpha \in Q_{t,n}$ is the set complementary to β in $\{1, \cdots, n\}$. Thus the complete coefficient of λ^{n-t} is

$$(-1)^t \sum_{\alpha \in Q_{t,n}} d(A[\alpha|\alpha]) = (-1)^t E_t(A).$$

2.14.2   If A is a 2-square matrix over **C**, then $E_1(A) = \text{tr }(A)$, $E_2(A) = d(A)$ and the characteristic polynomial is $p(\lambda) = \lambda^2 - \text{tr }(A)\lambda + d(A)$.

2.15   Characteristic roots

The n roots of the characteristic polynomial $p(\lambda) = d(\lambda I_n - A)$, each counted with its proper multiplicity, are called the *characteristic roots* of A. The set of characteristic roots of A will be denoted by $\lambda(A)$; and if all the characteristic roots are real, we will choose the notation so that

$$\lambda_1(A) \geq \lambda_2(A) \geq \cdots \geq \lambda_n(A). \tag{1}$$

The reason for the notation $E_t(A)$ is now clear: for, in general,

$$\begin{aligned} p(\lambda) = d(\lambda I_n - A) &= \prod_{j=1}^{n} (\lambda - \lambda_j(A)) \\ &= \sum_{t=0}^{n} \lambda^{n-t}(-1)^t E_t(\lambda_1(A), \cdots, \lambda_n(A)). \end{aligned} \tag{2}$$

Thus comparing **2.15(2)** and **2.13(1)** we see that

$$E_t(A) = E_t(\lambda_1(A), \cdots, \lambda_n(A)). \tag{3}$$

In other words, the sum of all the t-square principal subdeterminants of A is just the tth elementary symmetric function of the characteristic roots of A. Some important consequences of this are

$$\text{tr}(A) = \sum_{i=1}^{n} a_{ii} = \sum_{i=1}^{n} \lambda_i(A); \tag{4}$$

$$d(A) = E_n(\lambda_1(A), \cdots, \lambda_n(A)) = \prod_{i=1}^{n} \lambda_i(A). \tag{5}$$

The numbers $\lambda_1(A), \cdots, \lambda_n(A)$ go under several names: *eigenvalues, secular values, latent roots, proper values.*

Some elementary properties of characteristic roots are listed.

2.15.1 If $A = \begin{pmatrix} B & E \\ 0_{q,p} & D \end{pmatrix} \in M_n(\mathbf{C})$ where $B \in M_p(\mathbf{C})$ and $D \in M_q(\mathbf{C})$,

then the n characteristic roots of A are the p characteristic roots of B taken together with the q characteristic roots of D:

$$\lambda(A) = \{\lambda_1(B), \cdots, \lambda_p(B), \lambda_1(D), \cdots, \lambda_q(D)\}.$$

2.15.2 If $A = \begin{pmatrix} a_{11} & & \cdots & & a_{1n} \\ 0 & a_{22} & \cdots & & a_{2n} \\ \cdot & & & & \cdot \\ \cdot & & & & \cdot \\ \cdot & & & & \cdot \\ 0 & & \cdots & 0 & a_{nn} \end{pmatrix} \in M_n(\mathbf{C})$

in which the entries below the main diagonal are 0, then the characteristic roots of A are $a_{11}, \cdots, a_{nn}$, the main diagonal entries of A.

2.15.3 If $a \in \mathbf{C}$ and $A \in M_n(\mathbf{C})$, then $aI_n + A$ has characteristic roots $a + \lambda_i(A)$, $(i = 1, \cdots, n)$, and aA has characteristic roots $a\lambda_i(A)$, $(i = 1, \cdots, n)$. Thus we can write suggestively (if not properly) that $\lambda(aI_n + bA) = a + b\lambda(A)$ whenever a and b are in $\mathbf{C}$.

2.15.4 If $S \in M_n(\mathbf{C})$ and S is nonsingular, then $\lambda(S^{-1}AS) = \lambda(A)$ for any $A \in M_n(\mathbf{C})$.

2.15.5 If $A \in M_n(\mathbf{C})$, then the characteristic roots of A^p are $\lambda_i^p(A)$, $(i = 1, \cdots, n)$, for any positive integer p.

2.15.6 If $A \in M_n(\mathbf{C})$, then A is nonsingular if and only if $\lambda_i(A) \neq 0$, $(i = 1, \cdots, n)$. This follows immediately from 2.4.5 and 2.15(5).

2.15.7 If $A \in M_n(\mathbf{C})$, A is nonsingular, and $p \neq 0$ is an integer, then the characteristic roots of A^p are $\lambda_i^p(A)$, $(i = 1, \cdots, n)$.

2.15.8 If $A \in M_n(\mathbf{C})$, $a_0, \cdots, a_m$ are numbers in $\mathbf{C}$, and $B = a_0 I_n + a_1 A + \cdots + a_m A^m$, then the characteristic roots of B are

$$a_0 + a_1 \lambda_i(A) + a_2 \lambda_i^2(A) + \cdots + a_m \lambda_i^m(A), \qquad (i = 1, \cdots, n).$$

2.15.9 If $A \in M_n(\mathbf{C})$, $a_0, \cdots, a_m$ are numbers in $\mathbf{C}$, and $a_0 I_n + a_1 A + \cdots + a_m A^m = 0_{n,n}$, then any characteristic root r of A must satisfy the equation

$$a_0 + a_1 r + a_2 r^2 + \cdots + a_m r^m = 0.$$

2.15.10 If $A_p \in M_{n_p}(\mathbf{C})$, $(p = 1, \cdots, m)$, and n_p are positive integers, then $\lambda \left(\sum_{p=1}^{m} A_p \right) = \bigcup_{p=1}^{m} \lambda(A_p)$. In other words, the characteristic roots of a direct sum of matrices is just the set of all of the characteristic roots of the individual summands taken together.

2.15.11 If $A \in M_p(\mathbf{C})$ and $B \in M_q(\mathbf{C})$, then the characteristic roots of the Kronecker product $A \otimes B \in M_{pq}(\mathbf{C})$ are the pq numbers $\lambda_i(A)\lambda_j(B)$, $(i = 1, \cdots, p; j = 1, \cdots, q)$.

2.15.12 If $A \in M_n(\mathbf{C})$ and $1 \leq k \leq n$, then the characteristic roots of $C_k(A)$ are the $\binom{n}{k}$ products $\lambda_{i_1}(A) \cdots \lambda_{i_k}(A)$ where
$$1 \leq i_1 < i_2 < \cdots < i_k \leq n.$$

In the notation of **2.1**, $\lambda_\omega(C_k(A)) = \lambda_\omega(A) = \prod_{t=1}^{k} \lambda_{i_t}(A)$ for $\omega = (i_1, \cdots, i_k) \in Q_{k,n}$; thus
$$\operatorname{tr}(C_k(A)) = E_k(\lambda_1(A), \cdots, \lambda_n(A)) = \sum_{\omega \in Q_{k,n}} \lambda_\omega(A).$$

2.15.13 If $A \in M_n(\mathbf{C})$, then the characteristic roots of A^* are the complex conjugates of the characteristic roots of A. Thus $\lambda_i(A^*) = \overline{\lambda_i(A)}$, $(i = 1, \cdots, n)$; $\lambda(A^T) = \lambda(A)$, $\lambda_i(\overline{A}) = \overline{\lambda_i(A)}$, $(i = 1, \cdots, n)$.

2.15.14 If $A \in M_n(\mathbf{C})$ and $1 \leq k \leq n$, then the characteristic roots of $P_k(A)$ are the $\binom{n + k - 1}{k}$ products $\lambda_{i_1}(A) \cdots \lambda_{i_k}(A)$ where
$$1 \leq i_1 \leq i_2 \leq \cdots \leq i_k \leq n.$$

In the notation of **2.1**,
$$\lambda_\omega(P_k(A)) = \lambda_\omega(A) = \prod_{t=1}^{k} \lambda_{i_t}(A)$$
for $\omega = (i_1, \cdots, i_k) \in G_{k,n}$; thus
$$\operatorname{tr}(P_k(A)) = \sum_{\omega \in G_{k,n}} \lambda_\omega(A).$$

This latter function is called the *completely symmetric* function of the numbers $\lambda_1(A), \cdots, \lambda_n(A)$ and is usually designated by $h_k(\lambda_1(A), \cdots, \lambda_n(A))$.

2.15.15 If $A \in M_{m,n}(\mathbf{C})$, $B \in M_{n,m}(\mathbf{C})$, $(n \geq m)$, and $\varphi(\lambda) = d(\lambda I_m - AB)$, $\theta(\lambda) = d(\lambda I_n - BA)$, then $\theta(\lambda) = \lambda^{n-m}\varphi(\lambda)$. This implies that the nonzero characteristic roots of AB and BA are the same.

2.15.16 If $A_t \in M_n(\mathbf{C})$, $(t = 1, \cdots, k)$, and $A_iA_j = A_jA_i$, $(i, j = 1, \cdots, k)$, and $p(\lambda_1, \lambda_2, \cdots, \lambda_k)$ is a polynomial with complex coefficients in the indeterminates $\lambda_1, \cdots, \lambda_k$, then the characteristic roots of $B = p(A_1, A_2, \cdots, A_k)$ are the numbers $\beta_j = p(\alpha_{1j}, \alpha_{2j}, \cdots, \alpha_{kj})$, $(j = 1, \cdots, n)$, where α_{tj}, $(j = 1, \cdots, n)$, are the characteristic roots of A_t in some fixed order. For example, if A_1 and A_2 commute, then the characteristic roots of $A_1 A_2$ are $\alpha_{11}\alpha_{21}$, $\alpha_{12}\alpha_{22}$, $\cdots$, $\alpha_{1n}\alpha_{2n}$ where $\alpha_{11}, \cdots, \alpha_{1n}$ and $\alpha_{21}, \cdots, \alpha_{2n}$ are the characteristic roots of A_1 and A_2 in some order.

2.15.17 A somewhat more general result along these lines is the *Frobenius theorem*: If A and B are n-square matrices over **C** and both A and B commute with the commutator $AB - BA$, and if $f(\lambda_1, \lambda_2)$ is any polynomial in the indeterminates λ_1, λ_2 with complex coefficients, then there is an ordering of the characteristic roots of A and B, α_i, β_i, $(i = 1, \cdots, n)$, such that the characteristic roots of $f(A, B)$ are $f(\alpha_i, \beta_i)$, $(i = 1, \cdots, n)$.

2.15.18 If $A \in M_n(\mathbf{C})$ and A is nilpotent (i.e., $A^r = 0_{n,n}$ for some positive integer r), then $\lambda_i(A) = 0$, $(i = 1, \cdots, n)$. This is a consequence of **2.15.5**.

2.16 Examples

2.16.1 Let A be the (v,k,λ)-matrix in **2.4.10**. According to **2.15.3** the characteristic roots of AA^T are of the form $(k - \lambda) + \lambda\lambda_t(J)$. The characteristic polynomial of J is easily computed to be $x^{v-1}(x - v)$ and hence the characteristic roots of AA^T are $k + \lambda(v - 1)$ with multiplicity 1 and $k - \lambda$ with multiplicity $v - 1$. According to **2.15(5)** and **2.4.1** $d^2(A) = (k - \lambda)^{v-1}(k + \lambda(v - 1))$. But it was also proved in **2.4.10** that

$$\lambda = k(k - 1)/(v - 1)$$

and hence $d^2(A) = (k - \lambda)^{v-1}k^2$. Thus the formula for the adjoint of A given in **2.4.10** simplifies to $B = \pm(k - \lambda)^{(v-3)/2}(kA^T - \lambda J)$.

2.16.2 Let $P \in M_n(\mathbf{C})$ be the permutation matrix

$$P = (P_{ij}) = \begin{pmatrix} 0 & 1 & 0 & \cdots & 0 & 0 \\ 0 & 0 & 1 & \cdots & 0 & 0 \\ \vdots & & & \ddots & & \vdots \\ & & & & & \\ 0 & 0 & 0 & \cdots & 0 & 1 \\ 1 & 0 & 0 & \cdots & 0 & 0 \end{pmatrix} = (\delta_{i+1,j}), \qquad (\delta_{n+1,j} = \delta_{1j}).$$

Let $S \in M_n(\mathbf{C})$ be the matrix defined by $S_{pk} = n^{-1/2}\theta^{k(n-p)}$, $(k = 1, \cdots, n; p = 1, \cdots, n)$, where $\theta = e^{i2\pi/n} = \cos(2\pi/n) + i\sin(2\pi/n)$. We show that S is nonsingular, that $S^{-1} = S^*$, and that

$$S^{-1}PS = \text{diag}(\theta^{n-1}, \theta^{n-2}, \cdots, \theta, 1).$$

Thus, according to **2.15.2** and **2.15.4**, $1, \theta, \theta^2, \cdots, \theta^{n-1}$ are the characteristic roots of P. Now

$$(S^*S)_{pk} = \sum_{j=1}^{n} S^*_{pj}S_{jk} = \sum_{j=1}^{n} \overline{S}_{jp}S_{jk}$$

$$= n^{-1} \sum_{j=1}^{n} \theta^{(k-p)(n-j)}$$

$$= \begin{cases} 1 \text{ if } k = p \\ 0 \text{ if } k \neq p. \end{cases}$$

Thus $S^*S = I_n$, so $S^* = S^{-1}$. We compute

$$(S^{-1}PS)_{pk} = (S^*PS)_{pk} = \sum_{j=1,t=1}^{n} S^*_{pj}P_{jt}S_{tk}$$

$$= \sum_{j=1}^{n} \overline{S}_{jp}S_{j+1,k} = n^{-1}\theta^{n-k} \sum_{j=1}^{k} \theta^{(p-k)j} = \theta^{n-k}\delta_{pk}.$$

Hence $S^{-1}PS = \text{diag}\,(\theta^{n-1}, \theta^{n-2}, \cdots, 1)$ as asserted. Notice that $(S^{-1}PS)^* = \text{diag}\,(\theta, \theta^2, \theta^3, \cdots, \theta^{n-1}, 1)$.

2.16.3 In the analysis of the Ising model for ferromagnetic materials, an interesting matrix appears: it is the product of the two $2n$-square matrices K and L

$$K = \sum_{i=1}^{n} \cdot A, \qquad L = \begin{pmatrix} \cos\beta & 0 & 0 & \cdots & 0 & -\sin\beta \\ \cdot & B & & & & \cdot \\ \cdot & & B & 0 & & \cdot \\ 0 & & & \cdot & & 0 \\ \cdot & 0 & & & \cdot & \cdot \\ \cdot & & & & B & \cdot \\ \sin\beta & 0 & 0 & \cdots & 0 & \cos\beta \end{pmatrix}$$

where $\qquad A = \begin{pmatrix} \cos\alpha & \sin\alpha \\ -\sin\alpha & \cos\alpha \end{pmatrix}, \quad B = \begin{pmatrix} \cos\beta & \sin\beta \\ -\sin\beta & \cos\beta \end{pmatrix}.$

Set $E_1 = \begin{pmatrix} 0 & 1 \\ 0 & 0 \end{pmatrix}$ and $E_2 = \begin{pmatrix} 0 & 0 \\ 1 & 0 \end{pmatrix}$ and let $Q \in M_{2n}(\mathbb{C})$ be the permutation matrix discussed in **2.16.2** with n replaced by $2n$. That is,

$$Q = (\delta_{i+1,j}), \qquad (i, j = 1, \cdots, 2n).$$

Observe that if P is the n-square matrix in **2.16.2**, then

$$Q = I_n \otimes E_1 + P \otimes E_2,$$

and, using **1.9.1**,

$$Q^{-1} = Q^T = I_n \otimes E_2 + P^{-1} \otimes E_1.$$

One verifies by direct multiplication using **1.9(1)** that

$$L = Q(I_n \otimes B)Q^{-1}.$$

Moreover, $K = I_n \otimes A$. And by repeated use of **1.9(1)** it is simple to compute that

$$T = KL = I_n \otimes A \cos \beta + P \otimes A E_2 \sin \beta + P^{-1} \otimes A E_1(-\sin \beta).$$

According to **2.16.2** and **2.15.4**,

$$
\begin{aligned}
\lambda(T) &= \lambda((S^{-1} \otimes I_2) T (S \otimes I_2)) \\
&= \lambda((I_n \otimes A \cos \beta) + (\text{diag}(\theta^{n-1}, \theta^{n-2}, \cdots, 1) \otimes A E_2 \sin \beta) \\
&\qquad + (\text{diag}(\theta, \theta^2, \cdots, \theta^{n-1}, 1) \otimes A E_1(-\sin \beta))) \\
&= \lambda\left(\sum_{t=1}^{n} \cdot A \begin{pmatrix} \cos \beta & -\theta^t \sin \beta \\ \theta^{n-t} \sin \beta & \cos \beta \end{pmatrix} \right).
\end{aligned}
$$

Thus by **2.15.10** the characteristic roots of T are the $2n$ numbers that are obtained by computing the characteristic roots of each of the n 2-square matrices

$$A \begin{pmatrix} \cos \beta & -\theta^t \sin \beta \\ \theta^{n-t} \sin \beta & \cos \beta \end{pmatrix}, \qquad (t = 1, \cdots, n).$$

2.17 Rank

If $A \in M_{m,n}(\mathbf{F})$, then the rank of A is the size of the largest nonvanishing subdeterminant of A. The rank of A will be denoted by $\rho(A)$. If $A = 0_{m,n}$, then $\rho(A) = 0$, otherwise $\rho(A) \geq 1$. The most important properties of rank are listed here.

2.17.1 (Frobenius inequality) If P, Q, and R are rectangular matrices and the product PQR is defined, then

$$\rho(PQ) + \rho(QR) \leq \rho(Q) + \rho(PQR). \tag{1}$$

Taking first R and then P to be the appropriate size zero matrices it follows from **2.17(1)** that

$$\rho(PQ) \leq \min \{\rho(P), \rho(Q)\}. \tag{2}$$

2.17.2 If $P \in M_{m,n}(\mathbf{F})$, then

$$\rho(P) = \rho(PP^*) = \rho(P^*P). \tag{3}$$

2.17.3 If $A \in M_n(\mathbf{F})$ and A is nonsingular, then **2.4.5** states that $d(A) \neq 0$, so $\rho(A) = n$. More generally, $\rho(A)$ is at least the number of nonzero characteristic roots of A.

It follows from **2.17(2)** that if P and Q are nonsingular, then

$$\rho(PAQ) = \rho(A). \tag{4}$$

For,

$$
\begin{aligned}
\rho(PAQ) \leq \min \{\rho(P), \rho(AQ)\} &= \min \{n, \rho(AQ)\} \\
&= \rho(AQ) \leq \min \{\rho(A), \rho(Q)\} = \rho(A) \\
&= \rho(P^{-1}(PAQ)Q^{-1}) \leq \rho(PAQ).
\end{aligned}
$$

2.17.4 $\rho(A) = \rho(A^T) = \rho(A^*).$

2.17.5 $\rho(A + B) \leq \rho(A) + \rho(B)$.

2.17.6   If

$$A = \begin{pmatrix} A_1 & * & \cdots & & * \\ 0 & A_2 & & & \cdot \\ \cdot & & \cdot & & \cdot \\ \cdot & & & \cdot & \cdot \\ \cdot & & & & * \\ 0 & \cdots & & 0 & A_k \end{pmatrix}$$

where A_i is a principal submatrix, $(i = 1, \cdots, k)$, then

$$\rho(A) \geq \sum_{i=1}^{k} \rho(A_i).$$

2.17.7   If $A = \sum_{i=1}^{k} {}^{\cdot} A_i$, then

$$\rho(A) = \sum_{i=1}^{k} \rho(A_i).$$

2.17.8   **(Sylvester's law)**   If $A \in M_{m,n}(\mathbf{F})$ and $B \in M_{n,q}(\mathbf{F})$, then

$$\rho(A) + \rho(B) - n \leq \rho(AB) \leq \min \{\rho(A), \rho(B)\}.$$

2.17.9   If $A \in M_{m,n}(\mathbf{F})$ and $B \in M_{p,q}(\mathbf{F})$, then

$$\rho(A \otimes B) = \rho(A)\rho(B).$$

2.17.10   If $A \in M_{m,n}(\mathbf{F})$ and $\rho(A) = k$, then

$$\rho(C_r(A)) = \binom{k}{r} \quad \text{and} \quad \rho(P_r(A)) = \binom{k + r - 1}{r}.$$

If $k < r$, then $\binom{k}{r} = 0$ which agrees with the fact that $C_r(A)$ has every entry 0.

2.18   Linear combinations

If $a_i = (a_{i1}, \cdots, a_{in})$, $(i = 1, \cdots, m)$, is a set of n-vectors and $c_1, \cdots, c_m$ is a set of elements of $\mathbf{F}$, then the vector $\sum_{i=1}^{m} c_i a_i$ is called a *linear combination* of $a_1, \cdots, a_m$. In case $\mathbf{F}$ is $\mathbf{R}$ or $\mathbf{C}$, the totality of linear combinations so obtained by allowing arbitrary choices for the numbers $c_1, \cdots, c_m$ is called the *vector space over* $\mathbf{F}$ *generated* or *spanned* by $a_1, \cdots, a_m$ and will be designated by $\langle a_1, \cdots, a_m \rangle$. In general, a subset U of $\mathbf{F}^n$ is a *vector space* if it has the property that any linear combination of vectors in U is also in U (this property is called *closure*). The intersection of two such vector spaces, written $U \cap V$, is the totality of vectors that U and V have in common. The *sum* of U and V, written $U + V$, is the totality of sums $u + v$, $u \in U$, and $v \in V$. It is clear from the definitions that both $U \cap V$ and $U + V$ are vector spaces (satisfy the closure condition).

2.19 Example

Take $\mathbf{F}$ to be $\mathbf{R}$, the set of real numbers, and let e_i be the n-tuple with 1 as ith coordinate and all other coordinates 0. Then $\langle e_1, \cdots, e_n \rangle$ is $\mathbf{R}^n$. For if $x = (x_1, \cdots, x_n)$, $x_i \in \mathbf{R}$, then $x = \sum_{i=1}^{n} x_i e_i$.

2.20 Linear dependence; dimension

If $a_i = (a_{i1}, \cdots, a_{in}) \in \mathbf{F}^n$, $(i = 1, \cdots, m)$, and $\sum_{i=1}^{m} c_i a_i = 0_n$ for some $c_i \in \mathbf{F}$, (not all $c_i = 0$), then $a_1, \cdots, a_m$ are said to be *linearly dependent*. Otherwise $a_1, \cdots, a_m$ are *linearly independent*. Suppose the space $U = \langle a_1, \cdots, a_m \rangle$ has k linearly independent vectors in it and no set of $k + 1$ vectors of U is linearly independent. Then U is said to have *dimension* k. This is written $\dim U = k$. Define $A \in M_{m,n}(\mathbf{F})$ to be the matrix whose ith row is a_i, $(i = 1, \cdots, m)$. Some properties of dimension follow.

2.20.1 $\rho(A) = \dim U$. In other words, $\rho(A)$ is the largest number of linearly independent rows of A.

2.20.2 If $\dim U = k$, then there exist k linearly independent vectors among the a_i, $(i = 1, \cdots, m)$, say $a_{i_1}, \cdots, a_{i_k}$, $(1 \leq i_1 < \cdots < i_k \leq m)$, such that $\langle a_{i_1}, \cdots, a_{i_k} \rangle = U$. The vectors $a_{i_1}, \cdots, a_{i_k}$ constitute a *basis* of U. In other words, any of the vectors in U and, in particular, any of $a_1, \cdots, a_m$ is a linear combination of the vectors $a_{i_1}, \cdots, a_{i_k}$.

2.20.3 If $m > n$, then $a_1, \cdots, a_m$ are always linearly dependent.

2.20.4 If $b_1, \cdots, b_n$ designate the column vectors of A (m-vectors) and $\rho(A) = k$, then $\dim \langle b_1, \cdots, b_n \rangle = k$. In other words, the largest number of linearly independent rows in a matrix is the same as the largest number of linearly independent columns (see **2.20.1**).

2.20.5 If $A \in M_{m,n}(\mathbf{F})$, $\rho(A) = k$, and rows numbered $i_1 < \cdots < i_k$ are linearly independent and columns numbered $j_1 < \cdots < j_k$ are linearly independent, then $d(A[i_1, \cdots, i_k | j_1, \cdots, j_k]) \neq 0$. In other words, the submatrix lying in the intersection of k linearly independent rows and k linearly independent columns is nonsingular.

2.20.6 If U and V are any two vector spaces of n-tuples, then

$$\dim (U + V) + \dim (U \cap V) = \dim U + \dim V. \tag{1}$$

If U is a subset of V, then $\dim U \leq \dim V$.

2.20.7 If $u_1, \cdots, u_m$ are vectors in a k-dimensional vector space and $m > k$, then $u_1, \cdots, u_m$ are linearly dependent.

2.20.8   The dimension of $\mathbf{R}^n$ is n. For, the vectors $e_1, \cdots, e_n$ in **2.19** are linearly independent. Clearly the dimension of $\mathbf{C}^n$ is also n.

2.20.9   The set $M_{m,n}(\mathbf{C})$ can be regarded as a vector space of mn-tuples by simply assigning some fixed order to the elements of every $A \in M_{m,n}(\mathbf{C})$, say the lexicographic one, so that A corresponds to the mn-tuple

$$(a_{11}, \cdots, a_{1n}, a_{21}, \cdots, a_{2n}, \cdots, a_{m1}, \cdots, a_{mn}).$$

Hence, by **2.20.7**, the dimension of $M_{m,n}(\mathbf{C})$ is mn. In fact, if $E_{p,q} \in M_{m,n}(\mathbf{C})$ is the matrix with 1 in the (p,q) position and 0's elsewhere, then the matrices $E_{i,j}$, $(i = 1, \cdots, m; j = 1, \cdots, n)$, form a basis for $M_{m,n}(\mathbf{C})$.

2.21   Example

Suppose that $U \subset \mathbf{C}^{mn}$ $(n \geq m)$ is a vector space of dimension $(n-1)m + 1$. Then there exists a matrix $B \in M_{m,n}(\mathbf{C})$ such that $\rho(B) = 1$ and

$$(b_{11}, \cdots, b_{1n}, \cdots, b_{m1}, \cdots, b_{mn}) \in U.$$

For, let V be the set of all mn-tuples which have 0 coordinates after the first n coordinates:

$$(c_{11}, \cdots, c_{1n}, 0, \cdots, 0).$$

Now the dimension of V is clearly n so that

$$
\begin{aligned}
\dim (U \cap V) &= \dim U + \dim V - \dim (U + V) \\
&= (n-1)m + 1 + n - \dim (U + V) \\
&\geq (n-1)m + 1 + n - nm \\
&= n - m + 1 \geq 1.
\end{aligned}
$$

Thus there must be a nonzero matrix $B \in M_{m,n}(\mathbf{C})$ which is zero outside the first row. Hence $\rho(B) = 1$.

3   Linear Equations and Canonical Forms

3.1   Introduction and notation

A system of linear equations over $\mathbf{F}$,

$$\sum_{j=1}^{n} a_{ij}x_j = b_i, \qquad (i = 1, \cdots, m), \tag{1}$$

in which $a_{ij} \in \mathbf{F}$ and $b_i \in \mathbf{F}$, $(i = 1, \cdots, m; j = 1, \cdots, n)$, and x_j, $(j = 1, \cdots, n)$, are the unknowns, can be abbreviated to

$$Ax = b \tag{2}$$

in which $A = (a_{ij}) \in M_{m,n}(\mathbf{F})$, x is the n-tuple $(x_1, \cdots, x_n)$, and b is the

m-tuple $(b_1, \cdots, b_m)$. The problem is: given the matrix A and the m-tuple b find all n-tuples x for which **3.1(2)** is true. From **1.4(1)** we can express **3.1(2)** as an equation involving the columns of A:

$$\sum_{j=1}^{n} x_j A^{(j)} = b. \tag{3}$$

The matrix in $M_{m,n+1}(\mathbf{F})$ whose first n columns are the columns of A in order and whose $(n + 1)$st column is the m-tuple b is called the *augmented matrix* of the system **3.1(1)** and is denoted by $[A:b]$. The system **3.1(1)** is said to be *homogeneous* whenever $b = 0_m$. In case $b \neq 0_m$, the system **3.1(1)** is said to be *nonhomogeneous*. If **3.1(2)** is nonhomogeneous, then $Ax = 0_m$ is called the *associated* homogeneous system. If the system is homogeneous, then $x = 0_n$ is called the *trivial* solution to **3.1(2)**. The essential facts about linear equations follow.

3.1.1 The system of equations **3.1(2)** has a solution, if and only if $b \in \langle A^{(1)}, \cdots, A^{(n)} \rangle$. For, $b \in \langle A^{(1)}, \cdots, A^{(n)} \rangle$, if and only if there exist $x_1, \cdots, x_n$ in $\mathbf{F}$ such that $\sum_{j=1}^{n} x_j A^{(j)} = b$.

3.1.2 The system **3.1(2)** has a solution, if and only if $\rho([A:b]) = \rho(A)$. For, $b \in \langle A^{(1)}, \cdots, A^{(n)} \rangle$, if and only if $\rho([A:b]) = \rho(A)$ (= number of linearly independent columns of A).

3.1.3 The set of solutions x of the homogeneous system $Ax = 0_m$ constitutes a vector space of n-tuples, called the *null space* of A. For, $A(cx + dy) = cAx + dAy$; and, if x and y are solutions, then $Ax = 0_m = Ay$ and $A(cx + dy) = 0_m$ for any constants c and d. The dimension of the null space of A is called the *nullity* of A and is designated by $\eta(A)$.

3.1.4 If z_0 is a particular solution to **3.1(2)**, then *any* solution z to **3.1(2)** is of the form $z = z_0 + y$ where y is a solution of the associated homogeneous system $Ax = 0_m$.

3.1.5 $\rho(A) + \eta(A) = n$. This says that the maximum number of linearly independent vectors in the null space of A is $n - \rho(A)$. Thus, if $\rho(A) = n$, then $\eta(A) = 0$ and $Ax = 0_m$ has only the trivial solution $x = 0_n$. In case $n = m$, then $Ax = 0_m$ has a nontrivial solution, if and only if $d(A) = 0$.

3.1.6 If $m = n$, the system **3.1(2)** has a unique solution, if and only if $d(A) \neq 0$.

3.1.7 To solve $Ax = b$

  (i)   find a particular solution z to $Ax = b$;
  (ii)   find $\eta(A) = n - \rho(A)$ linearly independent solutions $y^1, \cdots, y^{\eta(A)}$ to $Ax = 0_m$;
  (iii)   then any solution x of $Ax = b$ is of the form

$$x = \sum_{j=1}^{\eta(A)} c_j y^j + z$$

where $c_1, \cdots, c_{\eta(A)}$ are appropriate numbers in **F**.

The problems (i) and (ii) can be solved at the same time by means of the so-called *triangular* or *Hermite* or *row-echelon* form to be introduced shortly. Before doing this we define what is meant by elementary operations on a matrix.

3.2 Elementary operations

If **F** is either of the fields **R** or **C**, then the three *elementary row operations* on a matrix $A \in M_{m,n}(\mathbf{F})$ are:

Type I: interchanging rows;
Type II: adding a multiple of a row to another row;
Type III: multiplying a row by a nonzero number in **F**.

In case **F** happens to be the ring of integers **I** or the ring of polynomials in an indeterminate λ over **R** or over **C**, then the above type III operations are replaced by:

Type III′: multiplying a row by a *unit* of the ring **F**, i.e., by an element with a multiplicative inverse in **F**.

The units in **I** are the numbers ± 1 and the units in $\mathbf{C}[\lambda]$ or $\mathbf{R}[\lambda]$ are the nonzero elements of **C** or **R** respectively.

3.2.1 It is handy to have a shorthand notation for describing these operations:

$\mathrm{I}_{(i),(j)}$ means an interchange of rows i and j;
$\mathrm{II}_{(i)+c(j)}$ means that c times row j is added to row i;
$\mathrm{III}_{c(i)}$ means row i is multiplied by c.

3.3 Example

Consider

$$A = \begin{pmatrix} 2 & 3 & -1 & 1 \\ 3 & 2 & -2 & 2 \\ 5 & 0 & -4 & 4 \end{pmatrix}.$$

Then performing $\mathrm{II}_{(2)-1(1)}$, $\mathrm{II}_{(3)-2(1)}$, $\mathrm{I}_{(1),(2)}$, $\mathrm{II}_{(2)-2(1)}$, $\mathrm{II}_{(3)-1(1)}$, $\mathrm{II}_{(3)+1(2)}$ in succession produces the matrix

$$B = \begin{pmatrix} 1 & -1 & -1 & 1 \\ 0 & 5 & 1 & -1 \\ 0 & 0 & 0 & 0 \end{pmatrix}.$$

3.4 Elementary matrices

An elementary row operation on an $m \times n$ matrix A can be achieved by premultiplying A by an appropriate elementary matrix E. A type I elementary matrix $E_{(i),(j)}$ is the matrix that results by performing $\mathrm{I}_{(i),(j)}$ on the rows of the m-square identity matrix I_m. Similarly $\mathrm{II}_{(i)+c(j)}$ performed on the rows of I_m yields $E_{(i)+c(j)}$ and $\mathrm{III}_{c(i)}$ performed on the rows of I_m produces $E_{c(i)}$. Thus to interchange rows (i) and (j) in A forms the product $E_{(i),(j)}A$. For the matrices discussed in **3.3** one has

$$B = E_{(3)+1(2)}E_{(3)-1(1)}E_{(2)-2(1)}E_{(1),(2)}E_{(3)-2(1)}E_{(2)-1(1)}A.$$

3.4.1   The elementary matrices are nonsingular and

$$\begin{aligned}
E_{(i),(j)}^{-1} &= E_{(i),(j)}; \\
E_{(i)+c(j)}^{-1} &= E_{(i)-c(j)}; \\
E_{c(i)}^{-1} &= E_{c^{-1}(i)}.
\end{aligned}$$

3.4.2   If B results from A by a succession of elementary operations, then $\rho(B) = \rho(A)$; for $B = EA$ where E is a product of nonsingular elementary matrices. Moreover, the system of equations

$$Ax = b \tag{1}$$

has precisely the same set of solutions as the system

$$Bx = Eb. \tag{2}$$

The two systems **3.4(1)** and **3.4(2)** are said to be *equivalent*.

3.5 Example

Consider the system

$$\begin{aligned}
2x_1 + 3x_2 - x_3 + x_4 &= 3, \\
3x_1 + 2x_2 - 2x_3 + 2x_4 &= 4, \\
5x_1 + 0x_2 - 4x_3 + 4x_4 &= 6.
\end{aligned}$$

Then this can be written $Ax = b$ where A is the matrix in **3.3** and b is the triple $(3, 4, 6)$. Denote the product of the elementary matrices corresponding to the elementary operations in **3.3** by E and then $Ax = b$ is equivalent to the system $Bx = Eb$. It turns out that Eb is the triple $(1, 1, 0)$ so that the equivalent system has the form

$$\begin{aligned}
x_1 - x_2 - x_3 + x_4 &= 1, \\
0x_1 + 5x_2 + x_3 - x_4 &= 1, \\
0x_1 + 0x_2 + 0x_3 + 0x_4 &= 0.
\end{aligned}$$

Dropping the terms with 0 coefficients yields the system of two equations

$$x_1 - x_2 - x_3 + x_4 = 1,$$
$$5x_2 + x_3 - x_4 = 1.$$ (1)

The associated homogeneous system $Ax = 0_3$ thus has the same solutions as the system of two equations

$$x_1 - x_2 - x_3 + x_4 = 0,$$
$$5x_2 + x_3 - x_4 = 0.$$

Now solve for x_2 in terms of x_3 and x_4 in the last equation and substitute this in the first equation. Then

$$x_1 = \tfrac{4}{5}x_3 + \tfrac{4}{5}x_4,$$
$$x_2 = -\tfrac{1}{5}x_3 + \tfrac{1}{5}x_4.$$

Thus any 4-tuple in the null space of A is of the form

$$(\tfrac{4}{5}x_3 + \tfrac{4}{5}x_4, -\tfrac{1}{5}x_3 + \tfrac{1}{5}x_4, x_3, x_4) = x_3(\tfrac{4}{5}, -\tfrac{1}{5}, 1, 0) + x_4(\tfrac{4}{5}, \tfrac{1}{5}, 0, 1).$$

A basis for the null space of A is provided by the two 4-tuples $(4, -1, 5, 0)$ and $(4, 1, 0, 5)$ and $\eta(A) = 2$. To get any solution of the system we need one solution of 3.5(1). But let $x_3 = 1$, $x_4 = 5$ and compute that $x_2 = 1$, $x_1 = -3$. Hence according to 3.1.7 the solutions to the system 3.5(1) are the totality of 4-tuples of the form

$$x = c_1(4, -1, 5, 0) + c_2(4, 1, 0, 5) + (-3, 1, 1, 5)$$

for any choice of the constants c_1 and c_2.

The technique exhibited in the example 3.5 is perfectly general.

3.6 Hermite normal form

Suppose $A \in M_{m,n}(\mathbf{F})$ where $\mathbf{F}$ is either $\mathbf{R}$ or $\mathbf{C}$ and $\rho(A) = k$. Then there exists a sequence of elementary row operations that reduces A to a matrix $B \in M_{m,n}(\mathbf{F})$ of the following form:

(i)   the first k rows of B are nonzero; the rest are zero;
(ii) the first nonzero element in the ith row of B, $(i = 1, \cdots, k)$, is 1 and occurs in column n_i, $(n_1 < n_2 \cdots < n_k)$;
(iii) the only nonzero element in column n_i of B is the 1 in row i.

The rank of B $(= \rho(A))$ is the number of nonzero rows in B.

3.6.1   The reduction to Hermite form can be performed as follows:

Step 1.   Let n_1 denote the column number of the first nonzero column in A. By a type I operation bring a nonzero entry to the $(1, n_1)$ position and by a type III operation make this entry 1.

Step 2.   With type II operations make all the entries zero in column n_1 except the $(1,n_1)$ entry. Scan the resulting matrix for the next nonzero column having a nonzero element in rows $2, \cdots, m$, say column n_2, $(n_1 < n_2)$.

Step 3.   By type I operations involving only rows numbered $2, \cdots, m$ bring a nonzero element into the $(2,n_2)$ position, reduce it to 1 by a type III operation and then by type II operations make the elements in column n_2 zero except the 1 in the $(2,n_2)$ position. The entries in positions $(1,j)$, $(j = 1, \cdots, n_2 - 1)$, are unaltered by this last sequence of operations because the $(2,n_2)$ entry is the first nonzero entry in row 2 and $n_1 < n_2$.

The procedure clearly can be continued until the Hermite form is obtained.

3.7   Example

The matrix A considered in **3.3** was reduced there to

$$\begin{pmatrix} 1 & -1 & -1 & 1 \\ 0 & 5 & 1 & -1 \\ 0 & 0 & 0 & 0 \end{pmatrix}.$$

To complete the reduction of A to Hermite normal form multiply row 2 by $\frac{1}{5}$. Then add row 2 to row 1 to obtain the matrix

$$B = \begin{pmatrix} 1 & 0 & -\frac{4}{5} & \frac{4}{5} \\ 0 & 1 & \frac{1}{5} & -\frac{1}{5} \\ 0 & 0 & 0 & 0 \end{pmatrix}.$$

3.8   Use of the Hermite normal form in solving $Ax = b$

If B is the Hermite form for A and E is the product of elementary matrices for which $EA = B = (b_{ij})$, then the system equivalent to $Ax = b$ is given by **3.4(2)**; i.e., $Bx = Eb$.

(i)   If $\rho(B) = \rho(A) = r$ and for some $t = r + 1, \cdots, m, (Eb)_t \neq 0$, then the system has no solution; $Ax = b$ is said to be an *inconsistent* system.

(ii)   If $(Eb)_t = 0$, $(t = r + 1, \cdots, m)$, then the system does have solutions. The associated homogeneous system has precisely $\eta(B) = \eta(A) = n - r$ linearly independent solutions obtained as follows: B has the form

$$B = \begin{pmatrix}
0 & \cdots & 0 & 1 & * & \cdots & * & 0 & * & \cdots & * & 0 & * & \cdots & * & 0 & * & \cdots & & * \\
0 & \cdots & 0 & 0 & & \cdots & 0 & 1 & * & \cdots & * & 0 & * & \cdots & * & 0 & * & \cdots & & * \\
0 & \cdots & 0 & 0 & & \cdots & 0 & 0 & & \cdots & 0 & 1 & * & \cdots & * & 0 & * & \cdots & & * \\
\vdots & & & \vdots & & & & \vdots & & & & & & & & 0 & & & & \vdots \\
\vdots & & & \vdots & & & & \vdots & & & & & & & & & & & & \vdots \\
\vdots & & & \vdots & & & & \vdots & & & & & 0 & \cdots & 0 & 1 & * & \cdots & & * \\
\vdots & & & \vdots & & & & \vdots & & & & & & & & & & & & \vdots \\
0 & \cdots & 0 & 0 & & \cdots & & \vdots & & \cdots & & 0 & & \cdots & & 0 & & \cdots & & 0
\end{pmatrix}$$

We solve the tth equation for x_{n_t}, $(t = 1, \cdots, r)$, in terms of all the $n - r$ unknowns x_j where j is distinct from any of $n_1, \cdots, n_r$. To simplify notation let N be the set $\{n_1, \cdots, n_r\}$ and then this last statement is just

$$x_{n_t} = -\sum_{j \notin N} b_{tj} x_j.$$

Then

$$x = (x_1, \cdots, x_n)$$
$$= (x_1, \cdots, x_{n_1}, \cdots, x_{n_2}, \cdots, x_{n_r}, \cdots, x_n)$$
$$= (x_1, \cdots, -\sum_{j \notin N} b_{1j} x_j, \cdots, -\sum_{j \notin N} b_{2j} x_j, \cdots, -\sum_{j \notin N} b_{rj} x_j, \cdots, x_n).$$

This last vector can be written as a linear combination of $n - r$ vectors over $\mathbf{F}$, say $u_1, \cdots, u_{n-r}$ with x_j, $j \notin N$, as coefficients. These vectors $u_1, \cdots, u_{n-r}$ span the null space of A.

(iii)   A particular solution to $Bx = Eb$ can be obtained by assigning values in $\mathbf{F}$ to x_j, $(j \notin N)$, in $Bx = Eb$ and solving for x_j, $(j \in N)$.

3.9 Example

Assume that the system has been reduced to Hermite form

$$Bx = c$$

where

$$B = \begin{pmatrix}
0 & 1 & -2 & 0 & 6 & 0 & 2 & -3 \\
0 & 0 & 0 & 1 & 3 & 0 & 1 & 2 \\
0 & 0 & 0 & 0 & 0 & 1 & 5 & 3 \\
0 & 0 & 0 & 0 & 0 & 0 & 0 & 0
\end{pmatrix}$$

and $c = (1, 1, -1, 0)$. Then

$$x_2 = 2x_3 - 6x_5 - 2x_7 + 3x_8,$$
$$x_4 = -3x_5 - x_7 - 2x_8,$$
$$x_6 = -5x_7 - 3x_8$$

so that

$$x = (x_1, 2x_3 - 6x_5 - 2x_7 + 3x_8, x_3, -3x_5 - x_7 - 2x_8, x_5, -5x_7 - 3x_8, x_7, x_8)$$
$$= x_1(1, 0, 0, 0, 0, 0, 0, 0) + x_3(0, 2, 1, 0, 0, 0, 0, 0)$$
$$+ x_5(0, -6, 0, -3, 1, 0, 0, 0) + x_7(0, -2, 0, -1, -5, 1, 0)$$
$$+ x_8(0, 3, 0, -2, 0, -3, 0, 1).$$

A basis of the null space of B (and hence of A) is provided by the five vectors in the preceding sum. A particular solution z to $Bx = c$ is

$$z = (0, 1, 0, 1, 0, -1, 0, 0).$$

Thus the general solution to $Bx = c$ is the totality of vectors

$$x = z + \sum_{j=1}^{5} c_j u_j$$

where the c_j are arbitrary numbers.

If, on the other hand, the vector c happened to be $(1, 2, -1, 7)$, the system would be inconsistent.

3.10 Elementary column operations and matrices

An *elementary column operation* is one of the three types described in **3.2** with the single alteration of the word "row" to the word "column." We shall use the notation $I^{(i),(j)}$, $II^{(i)+c(j)}$, $III^{c(i)}$ to indicate the three types of column operations and similarly, $E^{(i),(j)}$, $E^{(i)+c(j)}$, $E^{c(i)}$ are the three types of elementary matrices. Any elementary column operation can be achieved by a corresponding post-multiplication by an elementary matrix analogously to **3.4**.

3.10.1 Suppose $A \in M_{m,n}(\mathbf{F})$ where $\mathbf{F}$ is $\mathbf{R}$ or $\mathbf{C}$ and $\rho(A) = k$. Then there exists a sequence of elementary row and column operations that reduce A to a matrix $D \in M_{m,n}(\mathbf{F})$

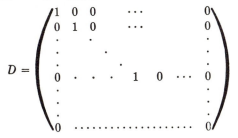

in which the only nonzero entries in D are 1's in positions (j,j), $(j = 1, \cdots, k)$.

This follows instantly from the Hermite form of A given in **3.6**. Simply eliminate the nonzero elements outside of columns $n_1, \cdots, n_k$ by type II column operations on B using the columns numbered $n_1, \cdots, n_k$. Follow this by type I column operations to put the 1's in the upper left block.

3.11 Examples

3.11.1   The matrix B in **3.9** is reduced to diagonal form by the succession of column operations: $\text{II}^{(7)-5(6)}$, $\text{II}^{(8)-3(6)}$, $\text{II}^{(5)-3(4)}$, $\text{II}^{(7)-1(4)}$, $\text{II}^{(8)-2(4)}$, $\text{II}^{(3)+2(2)}$, $\text{II}^{(5)-6(2)}$, $\text{II}^{(7)-2(2)}$, $\text{II}^{(8)+3(2)}$, $\text{I}^{(1),(2)}$, $\text{I}^{(2),(4)}$, $\text{I}^{(3),(6)}$.

3.11.2   If $a_i = (a_{i1}, \cdots, a_{in})$, $i = 1, \cdots, m$, and $m < n$, are linearly independent, then there exist vectors

$$a_t = (a_{t1}, \cdots, a_{tn}), \quad (t = m + 1, \cdots, n),$$

such that the vectors $a_1, \cdots, a_n$ are linearly independent. For, let A be the n-square matrix whose first m rows are $a_1, \cdots, a_m$ and whose remaining rows are all zero. Obtain nonsingular matrices P and Q such that

$$PAQ = \text{diag}\,(1, \cdots, 1, 0, \cdots, 0) = D,$$

the diagonal form described in **3.10.1**. The matrix P is a product of elementary matrices corresponding to elementary row operations on the first m rows of A. There are precisely m 1's along the main diagonal of D because $\rho(A) = m$. Now let $K = \text{diag}\,(0, \cdots, 0, 1, \cdots, 1)$ be an n-square diagonal matrix in which the first m positions along the main diagonal are 0 and the last $n - m$ are 1. Then $D + K = I_n$ so that

$$\begin{aligned}
A + P^{-1}KQ^{-1} &= P^{-1}DQ^{-1} + P^{-1}KQ^{-1} \\
&= P^{-1}(D + K)Q^{-1} \\
&= P^{-1}Q^{-1}.
\end{aligned}$$

Thus $\rho(A + P^{-1}KQ^{-1}) = n$ and the rows of $A + P^{-1}KQ^{-1}$ are linearly independent. However, P^{-1} is a product of elementary matrices corresponding to elementary row operations on the first m rows and hence P^{-1} has no effect on K: $P^{-1}KQ^{-1} = KQ^{-1}$. On the other hand, Q^{-1} is a product of elementary column matrices and hence KQ^{-1} has no nonzero entries among its first m rows. Thus the first m rows of $A + P^{-1}KQ^{-1}$ are just $a_1, \cdots, a_m$. Let $a_{m+1}, \cdots, a_n$ be the remaining rows and, as we have seen, all the vectors $a_1, \cdots, a_n$ are linearly independent.

3.11.3   If $A \in M_n(\mathbf{F})$, then A is nonsingular, if and only if, A is a product of elementary matrices. For, the diagonal form of A is the identity I_n and $A = PI_nQ$ where P and Q are products of elementary matrices.

3.12 Characteristic vectors

If $A \in M_n(\mathbf{C})$ and r is a characteristic root of A, then $d(A - rI_n) = 0$ and $A - rI_n$ is singular. The nullity $\eta(A - rI_n)$ is called the *geometric multiplicity* of r and any nonzero vector u in the null space of $A - rI_n$ is called a *characteristic vector (eigenvector, proper vector, latent vector) of A corresponding to r.* The *algebraic multiplicity* of r is just the number of times r occurs as a root of the characteristic polynomial $d(\lambda I_n - A)$.

Some general facts about characteristic vectors follow.

3.12.1   The algebraic multiplicity of a characteristic root r is always at least as much as the geometric multiplicity of r.

3.12.2   If $A \in M_n(\mathbf{C})$, $r_1, \cdots, r_k$ are distinct characteristic roots of A, and $u_1, \cdots, u_k$ are corresponding characteristic vectors, $Au_j = r_j u_j$, $(j = 1, \cdots, k)$, then $u_1, \cdots, u_k$ are linearly independent.

3.12.3   If $A \in M_n(\mathbf{C})$, then A has n linearly independent characteristic vectors $u_1, \cdots, u_n$, if and only if there exists a matrix U such that $U^{-1}AU$ is a diagonal matrix D. The columns of U are the vectors $u_1, \cdots, u_n$ and the main diagonal elements of D are the corresponding characteristic roots.

3.12.4   If $A \in M_n(\mathbf{C})$ and the n characteristic roots of A are distinct, then there exists a nonsingular matrix $U \in M_n(\mathbf{C})$ such that

$$U^{-1}AU = \text{diag}\,(\lambda_1(A), \cdots, \lambda_n(A)).$$

3.12.5   If $A \in M_n(\mathbf{C})$ and $r_1, \cdots, r_k$ are the distinct characteristic roots of A, then a necessary and sufficient condition for there to exist a nonsingular $U \in M_n(\mathbf{C})$ such that $U^{-1}AU$ is a diagonal matrix is that the algebraic multiplicity of r_t be equal to its geometric multiplicity, $(t = 1, \cdots, k)$.

3.13   Examples

3.13.1   The matrix A in **1.5.3** has $r = 0$ as root of algebraic multiplicity n. On the other hand, $\eta(A - 0I_n) = \eta(A) = 1$ and the geometric multiplicity of the characteristic root 0 is 1. A basis for the null space of A is the n-vector $e_1 = (1, 0, \cdots, 0)$.

3.13.2   The proof of the statement **3.12.1** affords an interesting application of the material in **3.11.2**. For suppose the geometric multiplicity of the characteristic root r is k, i.e., $\eta(rI_n - A) = k$. Let $u_1, \cdots, u_k$ be linearly independent n-vectors spanning the null space of $rI_n - A$. Then by **3.11.2** obtain $u_{k+1}, \cdots, u_n$ so that the whole set $u_1, \cdots, u_n$ is linearly independent. It follows that the matrix U whose columns are $u_1, \cdots, u_n$, in order, is nonsingular. Then by **1.4(3)**

$$(AU)^{(j)} = Au_j, \qquad (j = 1, \cdots, n),$$

and

$$Au_j = ru_j, \qquad (j = 1, \cdots, k).$$

Thus

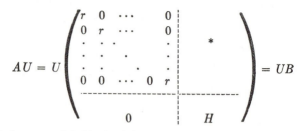

$$AU = U \begin{pmatrix} r & 0 & \cdots & & 0 & \vdots & \\ 0 & r & \cdots & & 0 & \vdots & \\ \cdot & \cdot & \cdot & & \cdot & \vdots & * \\ \cdot & \cdot & \cdot & & \cdot & \vdots & \\ \cdot & \cdot & & \cdot & \cdot & \vdots & \\ 0 & 0 & \cdots & 0 & r & \vdots & \\ \hline & & 0 & & & \vdots & H \end{pmatrix} = UB$$

in which the upper left block of the matrix B is a k-square diagonal matrix with r down its main diagonal and H is $(n - k)$-square. Now $U^{-1}AU = B$, $\lambda(A) = \lambda(U^{-1}AU) = \lambda(B)$ by 2.15.4. By 2.15.1, r is a characteristic root of B (and hence of A) of algebraic multiplicity at least k.

3.14   Conventions for polynomial and integral matrices

We proceed now to a consideration of matrices with polynomial or integral entries. The analysis for these two situations is the same and we shall introduce several standard terms from the general theory of rings in order to enable us to talk about them both at once. A *unit* in $\mathbf{I}$ is either of the numbers ± 1 and a *unit* in $\mathbf{F}[\lambda]$ is any nonzero element of $\mathbf{F}$ (here $\mathbf{F}$ is either of the fields $\mathbf{R}$ or $\mathbf{C}$). The *norm* of an integer p is just its absolute value and the *norm* of a polynomial $p(\lambda)$ is its degree. Both $\mathbf{I}$ and $\mathbf{F}[\lambda]$ have the property that given a and b, $(b \neq 0)$, there exist unique r and q such that

$$a = bq + r$$

in which the norm of r is less than the norm of b or $r = 0$; in the case of the ring of integers, r is taken to be nonnegative. We shall use the notation $b|a$, read "b divides a," in the case $r = 0$; if b does not divide a we write $b \nmid a$. The *greatest common divisor* of a set of nonzero integers $p_1, \cdots, p_m$ is just the largest positive integer p that divides them all. Similarly, the *greatest common divisor* of a set of nonzero polynomials $p_1(\lambda), \cdots, p_m(\lambda)$ is the polynomial of highest degree, $p(\lambda)$, that divides them all. We will always take this polynomial to be *monic*, i.e., the coefficient of the highest power of λ is 1. We write

$$p = \gcd (p_1, \cdots, p_m), \qquad p(\lambda) = \gcd (p_1(\lambda), \cdots, p_m(\lambda)).$$

The integers $p_1, \cdots, p_m$ are *relatively prime* if $p = 1$; and the polynomials $p_1(\lambda), \cdots, p_m(\lambda)$ are relatively prime, if $p(\lambda) = 1$. If a is a nonunit element in either ring not divisible by anything except a unit or a unit multiple of itself, then a is called a *prime*. In 3.15 to 3.24 we shall assume that the matrices considered are over either $\mathbf{F}[\lambda]$ or $\mathbf{I}$ unless it is stated otherwise.

For the purposes of these sections only, **K** will be used to designate either one of these rings when it is not necessary to distinguish between them.

3.15 Determinantal divisors

Let A be an $m \times n$ matrix and suppose $1 \leq k \leq \min \{m, n\}$. If A has at least one nonzero k-square subdeterminant, define f_k to be the greatest common divisor of all kth order subdeterminants of A. That is,

$$f_k = \gcd \{d(A[\omega|\varphi]), \; \omega \in Q_{k,m}, \; \varphi \in Q_{k,n}\},$$

if not all $d(A[\omega|\varphi])$ are zero; otherwise $f_k = 0$. The elements f_k are called the *determinantal divisors* of A. We set $f_0 = 1$ for notational convenience.

3.16 Examples

3.16.1   Let $A = \begin{pmatrix} 0 & 2 & 3 \\ 4 & 1 & 5 \\ 1 & 0 & 2 \end{pmatrix}$.

Then the sets of nonzero, 1-square, 2-square, and 3-square subdeterminants are respectively

$$\{2, 3, 4, 1, 5, 2\},$$
$$\{-8, -12, 7, -2, -3, 4, -1, 3, 2\},$$
$$\{-9\}.$$

Hence $f_1 = 1$, $f_2 = 1$, $f_3 = 9$.

3.16.2   Let A be the n-square matrix with r along the main diagonal, 1 in positions $(i, i+1)$, $(i = 1, \cdots, n-1)$, and 0 elsewhere. Rather than consider A itself we will compute the determinantal divisors of the characteristic matrix $\lambda I_n - A = B$ as a matrix over $\mathbf{F}[\lambda]$:

$$B = \begin{pmatrix} \lambda - r & -1 & 0 & \cdots & & 0 \\ 0 & \lambda - r & -1 & \cdots & & \cdot \\ \cdot & \cdot & \cdot & \cdot & & \cdot \\ \cdot & \cdot & & \cdot & \cdot & \cdot \\ \cdot & \cdot & & & \cdot & 0 \\ 0 & 0 & \cdots & & \lambda - r & -1 \\ 0 & 0 & \cdots & & 0 & \lambda - r \end{pmatrix}.$$

Now

$$d(B(n, 1, 2, \cdots, j-1|1, \cdots, j)) = (-1)^{n-i}, \qquad (j = 1, \cdots, n-1),$$
$$d(B) = (\lambda - r)^n.$$

Thus $f_1 = \cdots = f_{-1} = 1$ and $f_n = (\lambda - r)^n$.

3.17 Equivalence

3.17.1 If $A \in M_n(\mathbf{K})$, then A is called a *unit* matrix, if A^{-1} exists and $A^{-1} \in M_n(\mathbf{K})$.

3.17.2 If $A \in M_n(\mathbf{K})$, then A is a unit matrix, if and only if (a) $d(A)$ is a unit in $\mathbf{K}$, or (b) A is a product of elementary matrices.

3.17.3 If A and B are in $M_{m,n}(\mathbf{K})$, then A is *equivalent* to B over $\mathbf{K}$, if there exist unit matrices $P \in M_m(\mathbf{K})$ and $Q \in M_n(\mathbf{K})$ such that $A = PBQ$. It follows that equivalent matrices have the same rank.

3.17.4 A is equivalent to B over $\mathbf{K}$, if and only if A can be obtained from B by a sequence of elementary row and column operations.

The following result is of special importance.

3.17.5 Equivalent matrices have the same determinantal divisors. For if $A = PBQ$, then any kth order subdeterminant of A is a sum of multiples of kth order subdeterminants of B, by the Binet-Cauchy Theorem **2.4.14**. Hence the gcd of the kth order subdeterminants of B must divide all the kth order subdeterminants of A. But $B = P^{-1}AQ^{-1}$ and P^{-1} and Q^{-1} have their entries in $\mathbf{K}$—they are unit matrices (see **2.4.9**). The same argument shows that the kth determinantal divisor of A divides the kth determinantal divisor of B. Thus they are equal.

3.17.6 If $f_k = 0$, then $f_{k+1} = 0$. For if all k-square subdeterminants are zero, then all $(k + 1)$-square subdeterminants are 0 (use **2.4.8** to see this).

3.17.7 If f_r is not 0, then $f_r | f_{r+1}$ (use **2.4.8** again).

3.18 Example

Let $A \in M_v(\mathbf{I})$ be the (v,k,λ)-matrix discussed in **2.16.1**. Suppose that $k - \lambda > 1$, gcd $(k, k - \lambda) = 1$, and that $k - \lambda$ has no factor which is the square of a prime. Now $d(A) = \pm k(k - \lambda)^{(v-1)/2}$ and since A has only 0 and 1 entries, $d(A)$ is an integer. Hence $(v - 1)/2$ must be an integer because $k - \lambda$ is not a square. Thus v is odd and $f_v = k(k - \lambda)^{(v-1)/2}$. Using the formula for the adjoint $B = (b_{ij})$ given in **2.16.1** we have

$$b_{ij} = \pm (k - \lambda)^{(v-3)/2}(ka_{ji} - \lambda).$$

In case $a_{ji} = 0$, $b_{ij} = \pm\lambda(k - \lambda)^{(v-3)/2}$. In case $a_{ji} = 1$, $b_{ij} = \pm(k - \lambda)^{(v-1)/2}$. Both possibilities must occur in any (v,k,λ)-matrix, otherwise $k = v$ which contradicts $0 < \lambda < k < v$ (see **2.4.10**). The numbers b_{ij} are to within sign the $(v - 1)$-square subdeterminants of A. Thus $f_{v-1} = $ gcd $(\lambda(k - \lambda)^{(v-3)/2}, (k - \lambda)^{(v-1)/2})$. Since $(k - \lambda)$ and λ have no common factors, it follows that $f_{v-1} = (k - \lambda)^{(v-3)/2}$. Note that, in conformity with **3.17.7**, $f_{v-1} | f_v$.

We will be able to compute the rest of the determinantal divisors of A after we describe a few properties of the invariant factors to be defined in the next section.

3.19   Invariant factors

Let $A \in M_{m,n}(\mathbf{K})$ and suppose $f_0 = 1, f_1, \cdots, f_r$ are the nonzero determinantal divisors of A. According to 3 17.7, $f_{k-1}|f_k$, $(k = 1, \cdots, r)$. The quotients q_k defined by $f_k = q_k f_{k-1}$, $(k = 1, \cdots, r)$, are called the *invariant factors* of A. Note that $f_k = q_1 \cdots q_k$, $(k = 1, \cdots, r)$.

3.20   Elementary divisors

The invariant factors q_k, $(k = 1, \cdots, r)$, can be factored (uniquely except for order) into products of powers of primes $p_1, \cdots, p_m$:

$$q_k = p_1^{e_{k1}} p_2^{e_{k2}} \cdots p_m^{e_{km}}, \qquad (k = 1, \cdots, r). \tag{1}$$

The e_{kt} are nonnegative integers and $p_1, \cdots, p_m$ are all the distinct prime factors of the invariant factors. Each of the factors $p_t^{e_{kt}}$ for which $e_{kt} > 0$ is called an *elementary divisor* of A. An elementary divisor is counted once for each time it appears as a factor of an invariant factor. Moreover, in the case of a matrix A over $\mathbf{R}[\lambda]$ the elementary divisors will not be the same as if A were regarded as a matrix over $\mathbf{C}[\lambda]$. The elementary divisors of a matrix over $\mathbf{C}[\lambda]$ are always powers of linear polynomials.

3.20.1   If $C \in M_{m,n}(\mathbf{K})$, then C is equivalent to A, if and only if C and A have the same invariant factors (or determinantal divisors or elementary divisors). This result and **3.20.2** both are immediate consequences of the Smith normal form which we shall prove in **3.22.1**.

3.20.2   Given the list of all elementary divisors of $A \in M_{m,n}(\mathbf{K})$, $(\rho(A) = r)$, the list of invariant factors can be computed systematically as follows. Scan the list of elementary divisors for all the highest powers of the distinct primes that appear. Multiply these together to obtain q_r. Then scan the remaining elementary divisors for all the highest powers of the distinct primes that appear and multiply these together to obtain q_{r-1}. Continue this process until the list of elementary divisors is exhausted and suppose q_k is the last invariant factor so obtained. If $k > 1$, then set $q_1 = q_2 = \cdots = q_{k-1} = 1$.

3.21   Examples

3.21.1   If the list of elementary divisors of $A \in M_6(\mathbf{R}[\lambda])$ is $\lambda - 1$, $\lambda - 1, \lambda^2 + 1, \lambda^2 + 1$, and $\rho(A) = 6$, then the invariant factors of A are $q_6 = (\lambda - 1)(\lambda^2 + 1)$, $q_5 = (\lambda - 1)(\lambda^2 + 1)$, $q_4 = q_3 = q_2 = q_1 = 1$.

3.21.2   Let

$$A = \begin{pmatrix} 0 & 1 & 0 & & & \\ 0 & 0 & 1 & & 0 & \\ 1 & -1 & 1 & & & \\ \hline & & & 0 & 1 & 0 \\ & 0 & & 0 & 0 & 1 \\ & & & 1 & -1 & 1 \end{pmatrix}, \qquad B = \lambda I_6 - A \in M_6(\mathbf{K}).$$

Then the invariant factors of B are easily seen to be $q_6 = (\lambda - 1)(\lambda^2 + 1)$, $q_5 = (\lambda - 1)(\lambda^2 + 1)$, $q_4 = q_3 = q_2 = q_1 = 1$. The elementary divisors of B regarded as a matrix over $\mathbf{K} = \mathbf{R}[\lambda]$ are $\lambda - 1$, $\lambda - 1$, $\lambda^2 + 1$, $\lambda^2 + 1$; whereas regarding it as a matrix over $\mathbf{K} = \mathbf{C}[\lambda]$ the elementary divisors are $\lambda - 1$, $\lambda - 1$, $\lambda - i$, $\lambda - i$, $\lambda + i$, $\lambda + i$.

3.21.3   Let A be the matrix in **3.16.1**. The invariant factors of $A \in M_3(\mathbf{I})$ are $q_1 = f_1/f_0 = 1$, $q_2 = f_2/f_1 = 1$, $q_3 = f_3/f_2 = 9$. There is exactly one elementary divisor, 3^2.

The following is a result of fundamental importance.

3.22   Smith normal form

Suppose that $A \in M_{m,n}(\mathbf{K})$, $(\rho(A) = r)$. Then A has precisely $r + 1$ nonzero determinantal divisors $f_0, \cdots, f_r$. The matrix A is equivalent over $\mathbf{K}$ to a matrix $B = (b_{ij})$ in which $b_{ii} = q_i = f_i/f_{i-1}$, $(i = 1, \cdots, r)$, and $b_{ij} = 0$ otherwise. Moreover, $q_i | q_{i+1}$, $(i = 1, \cdots, r)$. The matrix B is called the Smith normal form of A.

3.22.1   The proof of **3.22** is actually a description of an algorithm for obtaining the matrix B. We assume that $A \neq 0_{m,n}$, otherwise it is already in Smith normal form.

Step 1.   Search A for a nonzero entry of least norm and bring it to the (1,1) position by row and column interchanges. By subtracting appropriate multiples of the first row from rows $2, \cdots, m$ obtain a matrix in which every entry in the first column below the (1,1) position is either 0 or has smaller norm than the (1,1) entry. Perform the analogous column operations to make every entry in the first row to the right of the (1,1) entry either 0 or of smaller norm than the (1,1) entry. Let G denote the matrix on hand at this point.

Step 2.   Search the first row and column of G for the nonzero entry of least norm, bring it to the (1,1) position and repeat the procedure of step 1 to decrease the norm of every entry in the first row and column outside the (1,1) position. At each stage we decrease the maximum norm of the elements in the first row and column by at least 1. Hence this process must terminate in a finite number of steps with a matrix A_1 equivalent to G (and hence equivalent to A) which is 0 in the first row and column outside the (1,1) position. Let a denote the (1,1) element of A_1.

Step 3. Suppose b is the (i,j) entry of A_1, $(i > 1, j > 1)$, and a/b. Add row i to row 1 of A_1 to put b in the $(1,j)$ position. Now $b = qa + r$ and $r \neq 0$ (otherwise $a|b$): subtract q times column 1 from column j which results in r appearing in the $(1,j)$ position. The matrix on hand at this point has an entry of smaller norm than a. With this matrix repeat steps 1 and 2 to obtain a matrix A_2 equivalent to A which is 0 in the first row and column outside the (1,1) position.

The process in step 3 is repeated with A_2 to obtain A_3 etc. We thereby obtain a sequence of matrices $A_1, A_2, A_3, A_4, A_5, \cdots, A_s$ in which the norms of the (1,1) entries are strictly decreasing, and in which row 1 and column 1 are 0 outside the (1,1) position. Moreover we go on from A_p to obtain A_{p+1} if and only if there is an entry in $A_p(1|1)$ not divisible by the (1,1) entry of A_p. Since the norms of the (1,1) entries are decreasing, the process must terminate with a matrix $C = (c_{ij})$ equivalent to A with the properties

(i) $c_{11}|c_{ij}$, $(i > 1, j > 1)$ (c_{11} may be a unit)
(ii) $c_{1j} = 0$, $(j = 2, \cdots, n)$,
      $c_{i1} = 0$, $(i = 2, \cdots, m)$.

Now the whole procedure can be repeated with the matrix C in which the elementary operations are performed on rows $2, \cdots, m$ and columns $2, \cdots, n$: the result will be a matrix D which is 0 in rows and columns 1 and 2 except for the (1,1) and (2,2) entries. Since c_{11} was a divisor of c_{ij}, $(i > 1, j > 1)$, it follows that c_{11} is a divisor of every element of D (this property isn't lost by elementary operations). Thus D looks like

$$D = \begin{pmatrix} c_{11} & 0 & \cdots & 0 \\ 0 & d & \cdots & 0 \\ \cdot & \cdot & & \\ \cdot & \cdot & L & \\ \cdot & \cdot & & \\ 0 & 0 & & \end{pmatrix}, \qquad (L \in M_{m-2,n-2}(\mathbf{K})),$$

and $c_{11}|d$, $c_{11}|l_{ij}$, $d|l_{ij}$, $(i = 1, \cdots, m - 2; j = 1, \cdots, n - 2)$. Clearly this process can be continued until we obtain a matrix $H \in M_{m,n}(\mathbf{K})$ in which $h_{ii}|h_{i+1,i+1}$, $h_{i+1,i+1} \neq 0$, $(i = 1, \cdots, p - 1)$, $h_{ij} = 0$ otherwise. Moreover, H is equivalent to A. By **3.17.3** and **3.17.5**, $\rho(H) = \rho(A)$ and H and A have the same determinantal divisors. Now $\rho(H) = p = \rho(A) = r$. Let $1 \leq k \leq r$ and observe that the only nonzero k-square subdeterminants of H are of the form $\prod_{t=1}^{k} h_{i_t i_t}$. The gcd of all such products is $\prod_{i=1}^{k} h_{ii}$ [recall $h_{ii}|h_{i+1,i+1}$, $(i = 1, \cdots, r - 1)$]. By definition of the invariant factors in **3.19**,

$$\prod_{i=1}^{k} h_{ii} = \prod_{i=1}^{k} q_i, \qquad (i = 1, \cdots, r).$$

Thus

$$h_{11} = q_1, \ h_{11}h_{22} = q_1q_2, \ \cdots, \ h_{11} \cdots h_{rr} = q_1 \cdots q_r;$$

so $h_{22} = q_2, \ \cdots, \ h_{rr} = q_r$. Hence $H = B$ and the argument is complete.

3.23 Example

This is a continuation of the computation of the determinantal divisors of the (v,k,λ)-matrix considered in 3.18. Let $q_1, \ \cdots, \ q_v$ be the invariant factors of A so that $f_k = q_1 \cdots q_k$, $(k = 1, \ \cdots, v)$, and $q_i|q_{i+1}$, $(i = 1, \ \cdots, v - 1)$. Now

$$q_v = \frac{f_v}{f_{v-1}} = \frac{k(k - \lambda)^{(v-1)/2}}{(k - \lambda)^{(v-3)/2}} = k(k - \lambda).$$

Moreover, $q_i|k(k - \lambda)$, $(i = 1, \ \cdots, v - 1)$, and $q_1 \cdots q_{v-1} = f_{v-1} = (k - \lambda)^{(v-3)/2}$. It follows that each q_i, $(i = 1, \ \cdots, v - 1)$, is a power of $(k - \lambda)$ and moreover must divide $k(k - \lambda)$ [in which gcd $(k, k - \lambda) = 1$]. Thus $q_i = 1$ or $q_i = k - \lambda$, $(i = 1, \ \cdots, v - 1)$. This fact together with $q_1 \cdots q_{v-1} = (k - \lambda)^{(v-3)/2}$ and $q_i|q_{i+1}$, $(i = 1, \ \cdots, v - 1)$, shows that the last $(v - 3)/2$ of the integers $q_1, \ \cdots, q_{v-1}$ must be $(k - \lambda)$ and the remaining $(v - 1) - (v - 3)/2 = (v + 1)/2$ of the q_i are 1. Thus the complete list of determinantal divisors is:

$$f_i = q_1 \cdots q_i = \begin{cases} 1, & \left(i = 1, \ \cdots, \ \dfrac{(v + 1)}{2}\right); \\[2mm] (k - \lambda)^{i-(v+1)/2}, & \left(i = \dfrac{(v + 1)}{2} + 1, \ \cdots, v - 1\right); \\[2mm] k(k - \lambda)^{(v-1)/2}, & (i = v). \end{cases}$$

It follows from 3.17.2 and 3.22 that there exist v-square matrices P and Q with integral elements satisfying $d(P) = \pm 1$, $d(Q) = \pm 1$, and $PAQ =$ diag $(1, \ \cdots, 1, k - \lambda, \ \cdots, k - \lambda, k(k - \lambda))$ in which 1 appears $(v + 1)/2$ times and $k - \lambda$ appears $(v - 3)/2$ times.

3.24 Similarity

Suppose that A and B are two n-square matrices over $\mathbf{F}$. We say A and B are *similar over* $\mathbf{F}$, if and only if there exists a nonsingular n-square matrix S over $\mathbf{F}$ such that $A = SBS^{-1}$. There is an intimate connection between the similarity of matrices over $\mathbf{F}$ and the equivalence of matrices over $\mathbf{F}[\lambda]$.

3.24.1 Let A and B be in $M_n(\mathbf{F})$ and suppose λ is an indeterminate over $\mathbf{F}$. Suppose that $\lambda I_n - A$ and $\lambda I_n - B$ are equivalent.

$$(\lambda I_n - A) = P(\lambda I_n - B)Q \tag{1}$$

where P and Q are n-square unit matrices over $\mathbf{F}[\lambda]$. Then there exists a matrix $S \in M_n(\mathbf{F})$ such that

$$A = SBS^{-1}, \tag{2}$$

i.e., A and B are similar over $\mathbf{F}$.

3.24.2 The matrix S is computed as follows. Q is a unit matrix and hence has an inverse R with polynomial entries which can be computed directly from 2.4.9. Now scan the polynomial entries of R for the entries of highest degree, say m, and write

$$R = R_m \lambda^m + R'.$$

Next scan the entries of R' for those of highest degree $m - 1$ (the coefficients may all be 0) and write

$$R = R_m \lambda^m + R_{m-1} \lambda^{m-1} + R''.$$

Continue this until there are no terms left:

$$R = R_m \lambda^m + R_{m-1} \lambda^{m-1} + \cdots + R_1 \lambda + R_0.$$

For example, if

$$R = \begin{pmatrix} \lambda^3 + 2 & \lambda^2 \\ -\lambda & 1 \end{pmatrix},$$

then

$$R = \begin{pmatrix} 1 & 0 \\ 0 & 0 \end{pmatrix} \lambda^3 + \begin{pmatrix} 0 & 1 \\ 0 & 0 \end{pmatrix} \lambda^2 + \begin{pmatrix} 0 & 0 \\ -1 & 0 \end{pmatrix} \lambda + \begin{pmatrix} 2 & 0 \\ 0 & 1 \end{pmatrix}.$$

The matrix S in **3.24(2)** is given by

$$S = R_m B^m + R_{m-1} B^{m-1} + \cdots + R_1 B + R_0.$$

A proof of **3.24.1** can be made now along the following lines: From $(\lambda I_n - A) = P(\lambda I_n - B)Q$ we write

$$P^{-1}(\lambda I_n - A) = (\lambda I_n - B)Q = Q\lambda - BQ. \tag{3}$$

If we replace the indeterminate λ on the right by the matrix A (exactly as was done above in determining S) in both sides of **3.24(3)**, we obtain $WA - BW = 0$. Here W is the matrix obtained by replacing λ on the right by A in Q. Let R be the inverse of Q, as before. In other words, $R_m \lambda^m Q + R_{m-1} \lambda^{m-1} Q + \cdots + R_1 \lambda Q + R_0 Q = I_n$. On the other hand, $\lambda^t Q = Q\lambda^t$, $(t = 0, \cdots, m)$, and hence $R_m Q\lambda^m + R_{m-1} Q\lambda^{m-1} + \cdots + R_1 Q\lambda + R_0 Q = I_n$. If we replace λ by A on the right in the last expression, we obtain

$$\sum_{t=0}^{m} R_t W A^t = I_n. \tag{4}$$

From $WA = BW$ we obtain in succession $WA^t = B^t W$, $(t = 0, \cdots, m)$, and hence **3.24(4)** becomes

$$\left(\sum_{t=0}^{m} R_t B^t \right) W = I_n.$$

On the other hand, $S = \sum_{t=0}^{m} R_t B^t$ and thus $SW = I_n$, $W = S^{-1}$, and finally $A = W^{-1}BW = SBS^{-1}$.

Conversely, if there exists some nonsingular matrix $T \in M_n(\mathbf{F})$ for which $A = TBT^{-1}$, then

$$\lambda I_n - A = \lambda I_n - TBT^{-1} = T(\lambda I_n - B)T^{-1}.$$

Now $d(T)$ is a nonzero element of $\mathbf{F}$ and, by **3.17.2**, T is a unit matrix in $\mathbf{F}[\lambda]$. Hence $\lambda I_n - A$ and $\lambda I_n - B$ are equivalent.

3.24.3   We sum up the above. Two matrices A and B in $M_n(\mathbf{F})$ are similar over $\mathbf{F}$, if and only if their characteristic matrices $\lambda I_n - A$ and $\lambda I_n - B$ are equivalent over $\mathbf{F}[\lambda]$; i.e., if and only if their characteristic matrices have the same invariant factors (see **3.20.1**). The invariant factors of the characteristic matrix of A are called the *similarity invariants* of A.

3.25 Examples

3.25.1   The matrices $A = \begin{pmatrix} 1 & 1 \\ 1 & 1 \end{pmatrix}$ and $B = \begin{pmatrix} 0 & 0 \\ 0 & 2 \end{pmatrix}$ are similar over the real numbers. For,

$$\lambda I_2 - A = \begin{pmatrix} \lambda - 1 & -1 \\ -1 & \lambda - 1 \end{pmatrix},$$

$$\lambda I_2 - B = \begin{pmatrix} \lambda & 0 \\ 0 & \lambda - 2 \end{pmatrix};$$

the determinantal divisors of $\lambda I_2 - A$ are $f_1 = \gcd(\lambda - 1, 1) = 1$, $f_2 = \lambda(\lambda - 2)$. The determinantal divisors of $\lambda I_2 - B$ are clearly $f_1' = \gcd(\lambda, \lambda - 2) = 1$, $f_2' = \lambda(\lambda - 2)$. Hence $f_1 = f_1'$, $f_2 = f_2'$ and, by **3.24.3**, A and B are similar over the real numbers. In fact, using **3.24.2**, we can find the matrix S such that $A = SBS^{-1}$. For, the matrices P and Q in **3.24(1)** may be taken to be

$$P = \begin{pmatrix} 1 & \lambda^2 - \lambda + 1 \\ -1 & -\lambda^2 + \lambda + 1 \end{pmatrix},$$

$$Q = \tfrac{1}{2} \begin{pmatrix} -\lambda^2 + 3\lambda - 1 & -\lambda^2 + 3\lambda - 3 \\ 1 & 1 \end{pmatrix}.$$

Then

$$Q^{-1} = R = \begin{pmatrix} 1 & \lambda^2 - 3\lambda + 3 \\ -1 & -\lambda^2 + 3\lambda - 1 \end{pmatrix}$$

$$= \begin{pmatrix} 0 & 1 \\ 0 & -1 \end{pmatrix} \lambda^2 + \begin{pmatrix} 0 & -3 \\ 0 & 3 \end{pmatrix} \lambda + \begin{pmatrix} 1 & 3 \\ -1 & -1 \end{pmatrix}$$

and hence

$$S = \begin{pmatrix} 0 & 1 \\ 0 & -1 \end{pmatrix} \begin{pmatrix} 0 & 0 \\ 0 & 2 \end{pmatrix}^2 + \begin{pmatrix} 0 & -3 \\ 0 & 3 \end{pmatrix} \begin{pmatrix} 0 & 0 \\ 0 & 2 \end{pmatrix} + \begin{pmatrix} 1 & 3 \\ -1 & -1 \end{pmatrix}$$

$$= \begin{pmatrix} 1 & 1 \\ -1 & 1 \end{pmatrix}.$$

Of course, S can be easily computed by other methods.

3.25.2   Let

$$A = \begin{pmatrix} 0 & 1 & 0 & 0 & 0 & 0 \\ 0 & 0 & 1 & 0 & 0 & 0 \\ 0 & 0 & 0 & 1 & 0 & 0 \\ 0 & 0 & 0 & 0 & 1 & 0 \\ 0 & 0 & 0 & 0 & 0 & 1 \\ -1 & 2 & -3 & 4 & -3 & 2 \end{pmatrix}.$$

The elementary divisors of $\lambda I_6 - A$, in which A is regarded as a matrix over $\mathbf{R}$, are $(\lambda - 1)^2$, $(\lambda^2 + 1)^2$. If A is regarded as a matrix over $\mathbf{C}$, then the elementary divisors of $\lambda I_6 - A$ are $(\lambda - 1)^2$, $(\lambda - i)^2$, $(\lambda + i)^2$. Consider the matrix

$$B = \begin{pmatrix} 0 & 1 \\ -1 & 2 \end{pmatrix} \dotplus \begin{pmatrix} 0 & 1 & 0 & 0 \\ 0 & 0 & 1 & 0 \\ 0 & 0 & 0 & 1 \\ -1 & 0 & -2 & 0 \end{pmatrix}.$$

The elementary divisors of $\lambda I_6 - B$ (over $\mathbf{R}$) are directly computed to be $(\lambda - 1)^2$, $(\lambda^2 + 1)^2$. Hence, by **3.20.1**, the matrices $\lambda I_6 - A$ and $\lambda I_6 - B$ are equivalent over $\mathbf{R}[\lambda]$. It follows by **3.24.3** that A and B are similar over $\mathbf{R}$.

3.26   Elementary divisors and similarity

3.26.1   Two matrices A and B in $M_n(\mathbf{F})$ are similar over $\mathbf{F}$ if and only if $\lambda I_n - A$ and $\lambda I_n - B$ have the same elementary divisors (see **3.20.1**, **3.24.3**).

3.26.2   If A and B are in $M_n(\mathbf{R})$, then A and B are similar over $\mathbf{C}$ if and only if they are similar over $\mathbf{R}$.

3.26.3   If

$$B = \sum_{i=1}^{m} \dotplus A_i, \quad (A_i \in M_{n_i}(\mathbf{F}); i = 1, \cdots, m),$$

$$n = \sum_{i=1}^{m} n_i,$$

then the set of elementary divisors of $\lambda I_n - B$ is the totality of elementary divisors of all the $\lambda I_{n_i} - A_i$ taken together, $(i = 1, \cdots, m)$. The proof of this statement is deferred to **3.29.5**.

3.26.4 If A_i is similar to B_i over **F**, $(i = 1, \cdots, m)$, then $\sum_{i=1}^{m} A_i$ is similar to $\sum_{i=1}^{m} B_i$ over **F**.

3.26.5 If A is in $M_n(\mathbf{F})$, then A is similar over **F** to A^T. For, $\lambda I_n - A^T = (\lambda I_n - A)^T$ and therefore A^T and A have the same similarity invariants.

3.27 Example

Consider the matrix B over **R** given in **3.25.2**. Then, as in **3.26.3**, $m = 2$,

$$A_1 = \begin{pmatrix} 0 & 1 \\ -1 & 2 \end{pmatrix},$$

and

$$A_2 = \begin{pmatrix} 0 & 1 & 0 & 0 \\ 0 & 0 & 1 & 0 \\ 0 & 0 & 0 & 1 \\ -1 & 0 & -2 & 0 \end{pmatrix}.$$

The elementary divisor of $\lambda I_2 - A_1$ is $(\lambda - 2)^2$ and the elementary divisor of $\lambda I_4 - A_2$ is $(\lambda^2 + 1)^2$. By **3.26.3** the elementary divisors of $\lambda I_6 - B$ are $(\lambda - 2)^2$ and $(\lambda^2 + 1)^2$.

3.28 Minimal polynomial

If $A \in M_n(\mathbf{F})$, then the monic polynomial $\varphi(\lambda) \in \mathbf{F}[\lambda]$ of least degree for which $\varphi(A) = 0_{n,n}$ is called the *minimal polynomial* of A. We give below the essential facts about the minimal polynomial and related notions.

3.28.1 The minimal polynomial of A divides any polynomial $f(\lambda)$ for which $f(A) = 0_{n,n}$.

3.28.2 (Cayley-Hamilton theorem) If $\psi(\lambda)$ is the characteristic polynomial of A, then $\psi(A) = 0_{n,n}$ and thus the minimal polynomial of A divides the characteristic polynomial; a matrix for which the minimal polynomial is equal to the characteristic polynomial is called *nonderogatory*, otherwise *derogatory*.

Proof: By **2.4.9** we have

$$(\lambda I_n - A) \text{ adj } (\lambda I_n - A) = \psi(\lambda) I_n.$$

Clearly adj $(\lambda I_n - A)$ is a matrix with polynomial entries of degree not exceeding $n - 1$. Let

$$\text{adj } (\lambda I_n - A) = B_{n-1}\lambda^{n-1} + \cdots + B_1\lambda + B_0,$$
$$(B_j \in M_n(\mathbf{F}); j = 0, \cdots, n - 1),$$

and

$$\psi(\lambda) = c_n\lambda^n + \cdots + c_1\lambda + c_0, \qquad (c_j \in \mathbf{F}; j = 1, \cdots, n).$$

Then

$$(\lambda I_n - A)(B_{n-1}\lambda^{n-1} + \cdots + B_1\lambda + B_0) = (c_n\lambda^n + \cdots + c_1\lambda + c_0)I_n.$$

Comparing coefficients, we obtain

$$
\begin{aligned}
B_{n-1} &= c_nI_n \\
B_{n-2} - AB_{n-1} &= c_{n-1}I_n \\
& \cdot \\
& \cdot \\
& \cdot \\
B_0 \quad - AB_1 &= c_1I_n \\
- AB_0 &= c_0I_n.
\end{aligned}
$$

Multiplying the first of these equalities by A^n, the second by A^{n-1}, the jth by A^{n-i+1}, and adding them together yields

$$0_{n,n} = c_nA^n + c_{n-1}A^{n-1} + \cdots + c_1A + c_0I_n = \psi(A).$$

3.28.3 The minimal polynomial of A is equal to its similarity invariant of highest degree.

Proof: Let $\varphi(\lambda)$ denote the minimal polynomial and let $q_n(\lambda)$ be the similarity invariant of highest degree. That is, $q_n(\lambda)$ is the invariant factor of highest degree of the characteristic matrix $\lambda I_n - A$. The entries of $Q(\lambda) = \text{adj } (\lambda I_n - A)$ are all the $(n - 1)$-square subdeterminants of $\lambda I_n - A$. Hence the $(n - 1)$st determinantal divisor of $\lambda I_n - A, f_{n-1}(\lambda)$, is just the gcd of all the entries of $Q(\lambda)$: $Q(\lambda) = f_{n-1}(\lambda)D(\lambda)$ where the entries of $D(\lambda)$ are relatively prime. If $f(\lambda)$ is the characteristic polynomial of A, then by definition, $f(\lambda) = q_n(\lambda)f_{n-1}(\lambda)$ and hence $f_{n-1}(\lambda)D(\lambda)(\lambda I_n - A) = Q(\lambda)(\lambda I_n - A) = f(\lambda)I_n = q_n(\lambda)f_{n-1}(\lambda)I_n$. Clearly $f_{n-1}(\lambda) \neq 0$ [e.g., $d((\lambda I_n - A)(1|1))$ is a polynomial with leading term λ^{n-1}] and hence

$$D(\lambda)(\lambda I_n - A) = q_n(\lambda)I_n.$$

If we express both sides of **3.28(1)** as polynomials with matrix coefficients (see **3.24.2**) and replace λ by A, then it follows that $q_n(A) = 0_{n,n}$. Hence by **3.28.1**, $\varphi(\lambda)|q_n(\lambda)$. We set $q_n(\lambda) = \varphi(\lambda)h(\lambda)$. From $\varphi(A) = 0_{n,n}$ we conclude by the division theorem for polynomials that $\varphi(\lambda)I_n = C(\lambda)(\lambda I_n - A)$ where $C(\lambda)$ is a polynomial in λ with matrix coefficients. Thus, from **3.28(1)**,

$$D(\lambda)(\lambda I_n - A) = q_n(\lambda)I_n = h(\lambda)\varphi(\lambda)I_n = h(\lambda)C(\lambda)(\lambda I_n - A).$$

It follows from the uniqueness of division that

$$D(\lambda) = h(\lambda)C(\lambda). \tag{2}$$

If we now regard $D(\lambda)$ and $C(\lambda)$ in **3.28(2)** as matrices with polynomial entries, we conclude that $h(\lambda)$ is a divisor of the entries of $D(\lambda)$. But these are relatively prime and hence $h(\lambda)$ is a unit. Since $q_n(\lambda)$ and $\varphi(\lambda)$ are both monic we obtain $h(\lambda) = 1$ and $\varphi(\lambda) = q_n(\lambda)$.

3.28.4 The minimal polynomial of $A \dotplus B$, $[A \in M_n(\mathbf{F}), B \in M_m(\mathbf{F})]$, is the least common multiple of the minimal polynomials of A and B. Recall that the least common multiple of two polynomials is the monic polynomial of least degree that both divide.

3.28.5 The matrix $A \in M_n(\mathbf{F})$ is nonderogatory (see **3.28.2**), if and only if the first $n - 1$ similarity invariants of A are 1.

3.29 Companion matrix

If $p(\lambda) \in \mathbf{F}[\lambda]$, $(p(\lambda) = \lambda^k - \sum_{j=1}^{k} a_{k-j}\lambda^{k-j})$, then the matrix

$$
\begin{pmatrix}
0 & 1 & 0 & \cdot & \cdot & \cdot & \cdot & \cdot & 0 \\
0 & 0 & 1 & 0 & \cdot & \cdot & \cdot & \cdot & 0 \\
\cdot & \cdot & 0 & \cdot & \cdot & \cdot & \cdot & \cdot & 0 \\
\cdot & \cdot & \cdot & \cdot & & & & & \cdot \\
\cdot & \cdot & \cdot & & \cdot & & & & \cdot \\
\cdot & \cdot & \cdot & & & \cdot & & & \cdot \\
0 & 0 & 0 & \cdot & \cdot & \cdot & \cdot & 0 & 1 \\
a_0 & a_1 & a_2 & \cdot & \cdot & \cdot & \cdot & a_{k-2} & a_{k-1}
\end{pmatrix} \in M_k(\mathbf{F})
$$

is called the *companion matrix* of the polynomial $p(\lambda)$ and is designated by $C(p(\lambda))$; if $p(\lambda) = \lambda - a_0$, then $C(p(\lambda)) = (a_0) \in M_1(\mathbf{F})$. The matrix $C(p(\lambda))$ is nonderogatory and its characteristic and minimal polynomials both equal $p(\lambda)$.

3.29.1 If $p(\lambda) = (\lambda - a)^k$, $(a \in \mathbf{F})$, then

$$
H(p(\lambda)) = \begin{pmatrix}
a & 1 & & & \\
 & a & 1 & & 0 \\
 & & \cdot & \cdot & \\
 & & & \cdot & \cdot & \\
 & 0 & & \cdot & 1 \\
 & & & & a
\end{pmatrix} \in M_k(\mathbf{F})
$$

is called the *hypercompanion matrix* of the polynomial $(\lambda - a)^k$; if $p(\lambda) = \lambda - a$, then $H(p(\lambda)) = (a) \in M_1(\mathbf{F})$. The only nonzero entries in $H(p(\lambda))$ occur on the main diagonal and immediately above it. The matrix $H(p(\lambda))$ is nonderogatory and the characteristic and minimal polynomials both equal $p(\lambda)$. The following are three results of great importance.

3.29.2   If $A \in M_n(\mathbf{F})$, then A is similar over $\mathbf{F}$ to the direct sum of the companion matrices of its nonunit similarity invariants.

Proof: Suppose $q_i(\lambda)$, $(i = 1, \cdots, n)$, are the similarity invariants of A:

$$q_1(\lambda) = \cdots = q_k(\lambda) = 1 \text{ and } \deg q_j(\lambda) = m_j \geq 1, (j = k + 1, \cdots, n);$$

$n = \sum\limits_{j=k+1}^{n} m_j$; and $q_j(\lambda)|q_{j+1}(\lambda)$, $(j = 1, \cdots, n - 1)$ (see **3.22**). Let $Q_i = C(q_i(\lambda))$, $(i = k + 1, \cdots, n)$, and since (see **3.29**) Q_i is nonderogatory with minimum polynomial equal to $q_i(\lambda)$, we conclude that $\lambda I_{m_i} - Q_i$ is equivalent over $\mathbf{F}[\lambda]$ to diag $(1, \cdots, 1, q_i(\lambda))$. It follows immediately that $\lambda I_n - \sum\limits_{i=k+1}^{n}\!\!\cdot\, Q_i = \sum\limits_{i=k+1}^{n}\!\!\cdot\, \lambda I_{m_i} - Q_i$ is equivalent over $\mathbf{F}[\lambda]$ to diag $(1, \cdots, 1, q_{k+1}(\lambda), \cdots, q_n(\lambda)) = \text{diag } (q_1(\lambda), \cdots, q_n(\lambda))$. It is obvious that the invariant factors of this last matrix are just the $q_i(\lambda)$, $(i = 1, \cdots, n)$, and hence $\lambda I_n - \sum\limits_{i=k+1}^{n}\!\!\cdot\, Q_i$ is equivalent over $\mathbf{F}[\lambda]$ to $\lambda I_n - A$. The result now follows from **3.24.1**.

3.29.3   (Frobenius, or rational, canonical form)   If $A \in M_n(\mathbf{F})$, then A is similar over $\mathbf{F}$ to the direct sum of the companion matrices of the elementary divisors of $\lambda I_n - A$.

Proof: Let $p_t(\lambda)$, $(t = 1, \cdots, r)$, be all the different primes that occur in the factorization into primes of the nonunit similarity invariants of A, $q_{k+1}(\lambda), \cdots, q_n(\lambda)$. Let $e_{i1}(\lambda), \cdots, e_{im_i}(\lambda)$ be the positive powers of the various $p_t(\lambda)$ that occur in factorization of $q_i(\lambda)$; i.e., $q_i(\lambda) = e_{i1}(\lambda) \cdots e_{im_i}(\lambda)$. Let $Q_i = C(q_i(\lambda))$, $(i = k + 1, \cdots, n)$, and set $\deg q_i(\lambda) = k_i$. We show first that $\lambda I_{k_i} - Q_i$ is equivalent over $\mathbf{F}[\lambda]$ to

$$\sum\limits_{t=1}^{m_i}\!\!\cdot\, \text{diag } (1, \cdots, 1, e_{it}(\lambda)). \tag{1}$$

For, the determinantal divisors of $C(q_i(\lambda))$ are just $f_1(\lambda) = \cdots = f_{k_i-1}(\lambda) = 1$ and $f_{k_i}(\lambda) = q_i(\lambda)$. On the other hand, the set of $(k_i - 1)$-square subdeterminants of the matrix in **3.29(1)** clearly contains each of the polynomials $\prod\limits_{t=1, t \neq r}^{m_i} e_{it}(\lambda) = d_r(\lambda)$, $(r = 1, \cdots, m_i)$. The $e_{it}(\lambda)$, $(t = 1, \cdots, m_i)$, are powers of *distinct* prime polynomials and hence the gcd of the $d_r(\lambda)$, $(r = 1, \cdots, m_i)$, is 1. It follows that the determinantal divisors of the matrix **3.29(1)** are also just $f_1(\lambda) = \cdots = f_{k_i-1}(\lambda) = 1$ and $f_{k_i}(\lambda) = q_i(\lambda)$. Hence, from **3.20.1**, $\lambda I_{k_i} - Q_i$ is equivalent over $\mathbf{F}[\lambda]$ to the matrix in **3.29(1)**. Next, let $c_{it} = \deg e_{it}(\lambda)$, $(t = 1, \cdots, m_i; i = k + 1, \cdots, n)$. Then diag $(1, \cdots, 1, e_{it}(\lambda))$ is equivalent over $\mathbf{F}[\lambda]$ to $\lambda I_{c_{it}} - C(e_{it}(\lambda))$ (they both obviously have the same determinantal divisors) and hence we conclude that $\lambda I_n - \sum\limits_{i=k+1}^{n}\!\!\cdot\, Q_i$ is equivalent over $\mathbf{F}[\lambda]$ to

$$\lambda I_n - \sum_{i=k+1}^{n} \cdot \sum_{t=1}^{m_i} \cdot C(e_{it}(\lambda)).$$

Hence from **3.24.1** it follows that $\sum_{i=k+1}^{n} \cdot Q_i$ is similar to $\sum_{i=k+1}^{n} \cdot \sum_{t=1}^{m_i} \cdot C(e_{it}(\lambda))$ and from **3.29.2** the result follows.

3.29.4 (Jordan normal form) If $A \in M_n(\mathbf{C})$, then the elementary divisors of $\lambda I_n - A$ are of the form $(\lambda - a)^k$, $(k > 0)$, and A is similar over $\mathbf{C}$ to the direct sum of the hypercompanion matrices of all the elementary divisors of $\lambda I_n - A$. This direct sum is called the Jordan normal form. Since the Jordan normal form is a triangular matrix, the numbers appearing along the main diagonal are the characteristic roots of A.

Proof: Each of the two matrices $H((\lambda - a)^k)$ and $C((\lambda - a)^k)$ is non-derogatory (see **3.29** and **3.29.1**). Moreover, each matrix has the polynomial $(\lambda - a)^k$ as both its minimal and characteristic polynomial. Hence the matrices $H((\lambda - a)^k)$ and $C((\lambda - a)^k)$ have precisely the same similarity invariants. It follows (see **3.29.2**) that they are similar. The result is now an immediate consequence of **3.29.3**.

3.29.5 (Elementary divisors of a direct sum) We are now in a position to prove **3.26.3**. We prove that if $A = B \dotplus C$ where B and C are p- and q-square respectively, then the elementary divisors of $\lambda I_n - A$ are the elementary divisors of $\lambda I_p - B$ together with those of $\lambda I_q - C$. Let $S = \{e_i(\lambda), i = 1, \cdots, m\}$, denote the totality of elementary divisors of $\lambda I_p - B$ and $\lambda I_q - C$; each of the $e_i(\lambda)$ is a power of a prime polynomial. Scan S for the highest powers of all the distinct prime polynomials and multiply these together to obtain a polynomial $q_n(\lambda)$. Delete from S all the $e_i(\lambda)$ used so far and scan what remains of S for all the highest powers of the distinct prime polynomials. Multiply these together to obtain $q_{n-1}(\lambda)$. Continue in this fashion until S is exhausted to obtain polynomials $q_{k+1}(\lambda), \cdots, q_n(\lambda)$ (this is the process described in **3.20.2**). It is clear from this method of construction that $q_{k+1}(\lambda)|q_{k+2}(\lambda)$. It is also clear that the sum of the degrees of the $q_j(\lambda)$, $(j = k + 1, \cdots, n)$, is just $p + q = n$. Let Q be the n-square direct sum of the companion matrices of the $q_j(\lambda)$, $(j = k + 1, \cdots, n)$. It is easy to compute that the similarity invariants of Q are the $q_j(\lambda)$, $(j = k + 1, \cdots, n)$, and hence the elementary divisors of $\lambda I_n - Q$ are just the polynomials in S. Thus, by **3.29.3**, Q is similar to the direct sum of the companion matrices of the polynomials in S. By **3.29.3**, B and C are each similar to the direct sum of the companion matrices of the elementary divisors of $\lambda I_p - B$ and $\lambda I_q - C$ respectively. Thus A is similar to the direct sum of the companion matrices of the polynomials in S. Hence $\lambda I_n - A$ and $\lambda I_n - Q$ have the same elementary divisors and the proof is complete.

3.30 Examples

3.30.1 The companion matrix of the polynomial $p(\lambda) = \lambda^3 - \lambda^2 + \lambda - 1$ is

$$C(p(\lambda)) = \begin{pmatrix} 0 & 1 & 0 \\ 0 & 0 & 1 \\ 1 & -1 & 1 \end{pmatrix}.$$

3.30.2 The companion matrix of

$$\begin{aligned} p(\lambda) &= \lambda^6 - 2\lambda^5 + 3\lambda^4 - 4\lambda^3 + 3\lambda^2 - 2\lambda + 1 \\ &= (\lambda - 1)^2(\lambda^2 + 1)^2 \end{aligned}$$

is

$$E = \begin{pmatrix} 0 & 1 & 0 & 0 & 0 & 0 \\ 0 & 0 & 1 & 0 & 0 & 0 \\ 0 & 0 & 0 & 1 & 0 & 0 \\ 0 & 0 & 0 & 0 & 1 & 0 \\ 0 & 0 & 0 & 0 & 0 & 1 \\ -1 & 2 & -3 & 4 & -3 & 2 \end{pmatrix} \in M_6(\mathbf{R}).$$

The only nonunit invariant factor of $\lambda I_6 - E$ (according to **3.29**, E is nonderogatory) is $p(\lambda)$. Hence E is already in the form given by **3.29.2**. The elementary divisors in $\mathbf{R}[\lambda]$ of $\lambda I_6 - E$ are $p_1(\lambda) = (\lambda - 1)^2 = \lambda^2 - 2\lambda + 1$ and $p_2(\lambda) = (\lambda^2 + 1)^2 = \lambda^4 + 2\lambda^2 + 1$. Thus

$$C(p_1(\lambda)) = \begin{pmatrix} 0 & 1 \\ -1 & 2 \end{pmatrix},$$

$$C(p_2(\lambda)) = \begin{pmatrix} 0 & 1 & 0 & 0 \\ 0 & 0 & 1 & 0 \\ 0 & 0 & 0 & 1 \\ -1 & 0 & -2 & 0 \end{pmatrix}.$$

The Frobenius normal form in **3.29.3** states that E is similar over $\mathbf{R}$ to the direct sum

$$C(p_1(\lambda)) \dotplus C(p_2(\lambda)) = \left(\begin{array}{cc|cccc} 0 & 1 & & & & \\ -1 & -2 & & & 0 & \\ \hline & & 0 & 1 & 0 & 0 \\ & 0 & 0 & 0 & 1 & 0 \\ & & 0 & 0 & 0 & 1 \\ & & -1 & 0 & -2 & 0 \end{array} \right).$$

If E is regarded as a matrix in $M_6(\mathbf{C})$, then the elementary divisors of E are

$$h_1(\lambda) = (\lambda - 1)^2, \; h_2(\lambda) = (\lambda - i)^2, \; h_3(\lambda) = (\lambda + i)^2.$$

The corresponding hypercompanion matrices are

$$H(h_1(\lambda)) = \begin{pmatrix} 1 & 1 \\ 0 & 1 \end{pmatrix},$$

$$H(h_2(\lambda)) = \begin{pmatrix} i & 1 \\ 0 & i \end{pmatrix},$$

$$H(h_3(\lambda)) = \begin{pmatrix} -i & 1 \\ 0 & -i \end{pmatrix}.$$

The result 3.29.4 states that E is similar over $\mathbf{C}$ to its Jordan normal form:

$$H(h_1(\lambda)) \dotplus H(h_2(\lambda)) \dotplus H(h_3(\lambda)) = \left(\begin{array}{cc|cc|cc} 1 & 1 & & & & \\ 0 & 1 & & 0 & & 0 \\ \hline & & i & 1 & & \\ & 0 & 0 & i & & 0 \\ \hline & & & & -i & 1 \\ & 0 & & 0 & 0 & -i \end{array} \right).$$

3.31 Irreducibility

If $A \in M_n(\mathbf{F})$ and there exists a nonsingular $S \in M_n(\mathbf{F})$ for which

$$SAS^{-1} = C \dotplus D$$

where $C \in M_p(\mathbf{F})$, $D \in M_q(\mathbf{F})$, $p + q = n$, then A is said to be *reducible* over $\mathbf{F}$. Otherwise, A is *irreducible* over $\mathbf{F}$.

3.31.1 If $A \in M_n(\mathbf{F})$, then A is irreducible over $\mathbf{F}$, if and only if A is nonderogatory and the characteristic polynomial $d(\lambda I_n - A)$ is a power of a prime polynomial. Thus the hypercompanion matrix $H((\lambda - a)^k)$ is always irreducible. This implies that the Jordan normal form is "best" in the sense that no individual block appearing along the main diagonal is similar over $\mathbf{F}$ to a direct sum of smaller blocks. Similarly, the blocks down the main diagonal in the Frobenius normal form in 3.29.3 are companion matrices of powers of prime polynomials (the elementary divisors of $\lambda I_n - A$) and are therefore irreducible.

3.32 Similarity to a diagonal matrix

A matrix $A \in M_n(\mathbf{F})$ is similar to a diagonal matrix over $\mathbf{F}$ under the following conditions:

3.32.1 If and only if all the elementary divisors of $\lambda I_n - A$ linear.

3.32.2   If and only if the minimal polynomial of A has distinct linear factors in $\mathbf{F}[\lambda]$.

3.32.3   If and only if A has a set of n linearly independent characteristic vectors (this requires, of course, that the characteristic roots of A lie in $\mathbf{F}$).

3.32.4   If $\mathbf{F} = \mathbf{C}$ and A has distinct characteristic roots.

3.32.5   If and only if $d(B) \neq 0$ where $B = (b_{ij}) \in M_p(\mathbf{F})$ and $b_{ij} = \operatorname{tr}(A^{i+j})$, $(i, j = 0, \cdots, p-1)$, and p is the degree of the minimal polynomial of A.

Proof of **3.32.1**.   If the elementary divisors of $\lambda I_n - A$ are linear, then the rational canonical form of A (see **3.29.3**) is diagonal. Conversely, if A is similar to a diagonal matrix D, then, by **3.26.1**, the matrices $\lambda I_n - A$ and $\lambda I_n - D$ have the same elementary divisors and, by **3.26.3**, the elementary divisors of $\lambda I_n - D$ are linear.

Proof of **3.32.2**.   The definition of elementary divisors (see **3.20**) implies that the elementary divisors of a matrix in $M_n(\mathbf{F}[\lambda])$ are linear if and only if its invariant factor of highest degree has distinct linear factors in $\mathbf{F}[\lambda]$. The result follows by **3.28.3** and **3.32.1**.

Proof of **3.32.3**.   Note that if $S \in M_n(\mathbf{F})$ and $D = \operatorname{diag}(\lambda_1, \cdots, \lambda_n)$, then $(SD)^{(j)} = \lambda_j S^{(j)}$. Now, suppose that $S^{-1}AS = D$. Then

$$AS = SD,$$
$$(AS)^{(j)} = (SD)^{(j)},$$
$$AS^{(j)} = \lambda_j S^{(j)},$$

and, since S is nonsingular, the vectors $S^{(1)}, \cdots, S^{(n)}$ are linearly independent characteristic vectors of A.

Conversely, if $x_1, \cdots, x_n$ are linearly independent characteristic vectors of A and S is the matrix whose jth column is x_j, then

$$Ax_j = \lambda_j x_j,$$
$$AS^{(j)} = \lambda_j S^{(j)},$$
$$(AS)^{(j)} = (SD)^{(j)},$$

where $D = \operatorname{diag}(\lambda_1, \cdots, \lambda_n)$. The column vectors of S are linearly independent, and therefore

$$S^{-1}AS = D = \operatorname{diag}(\lambda_1, \cdots, \lambda_n).$$

Proof of **3.32.4**.   We show that if $A \in M_n(\mathbf{C})$ has n distinct characteristic roots, $\lambda_1, \cdots, \lambda_n$, then the corresponding characteristic vectors, $x_1, \cdots, x_n$, are linearly independent. For, suppose that $x_1, \cdots, x_r$ are linearly independent while $x_1, \cdots, x_r, x_{r+1}$ are linearly dependent; i.e., there exist $c_1, \cdots, c_{r+1} \in \mathbf{F}$, not all 0, such that

$$c_1 x_1 + \cdots + c_r x_r + c_{r+1} x_{r+1} = 0. \tag{1}$$

Clearly $c_{r+1} \neq 0$ for otherwise **3.32(1)** would imply linear dependence of $x_1, \cdots, x_r$. Multiply **3.32(1)** by A to obtain

$$c_1 A x_1 + \cdots + c_r A x_r + c_{r+1} A x_{r+1} = 0;$$

i.e.,

$$c_1 \lambda_1 x_1 + \cdots + c_r \lambda_r x_r + c_{r+1} \lambda_{r+1} x_{r+1} = 0. \tag{2}$$

Eliminate x_{r+1} between 3.32(1) and 3.32(2) so that

$$c_1 (\lambda_1 - \lambda_{r+1}) x_1 + \cdots + c_r (\lambda_r - \lambda_{r+1}) x_r = 0.$$

Since $x_1, \cdots, x_r$ are linearly independent and $\lambda_j \neq \lambda_{r+1}$, $(j = 1, \cdots, r)$, we must have $c_1 = \cdots = c_r = 0$. But this implies that $x_{r+1} = 0$, a contradiction.

3.33 Examples

3.33.1 Let
$$p(\lambda) = \lambda^4 - 2\lambda^3 + 2\lambda^2 - 2\lambda + 1 \in \mathbf{C}[\lambda].$$
Then
$$A = C(p(\lambda)) = \begin{pmatrix} 0 & 1 & 0 & 0 \\ 0 & 0 & 1 & 0 \\ 0 & 0 & 0 & 1 \\ -1 & 2 & -2 & 2 \end{pmatrix}.$$

The only nonunit similarity invariant of A is $p(\lambda)$ itself which can be factored (over $\mathbf{C}[\lambda]$) into $(\lambda - 1)^2 (\lambda + i)(\lambda - i)$. Since $p(\lambda)$ is not a product of distinct linear factors, it follows from 3.32.2 that A is not similar to a diagonal matrix over $\mathbf{C}$. However A is similar over $\mathbf{C}$ to the direct sum of the hypercompanion matrices of the elementary divisors of $\lambda I_4 - A$: $(\lambda - 1)^2$, $\lambda + i$, $\lambda - i$. Thus A is similar over $\mathbf{C}$ to

$$\begin{pmatrix} 1 & 1 & & 0 \\ 0 & 1 & & \\ \hline & & -i & 0 \\ 0 & & 0 & i \end{pmatrix}.$$

3.33.2 Although a matrix $A \in M_n(\mathbf{C})$ is not necessarily similar to a diagonal matrix, it is similar to a matrix in which the off-diagonal elements are arbitrarily small. For, we can obtain a nonsingular S for which $S^{-1}AS$ is Jordan normal form:

$$S^{-1}AS = B = \begin{pmatrix} b_{11} & b_{12} & 0 & \cdots & & 0 \\ 0 & b_{22} & b_{23} & 0 & \cdots & 0 \\ \vdots & & \ddots & & & \vdots \\ \vdots & & & \ddots & b_{n-1,n-1} & b_{n-1,n} \\ 0 & \cdots & & & 0 & b_{n,n} \end{pmatrix}.$$

Let $E = \mathrm{diag}\,(1, \delta, \cdots, \delta^{n-1})$, $(\delta > 0)$, then

$$(SE)^{-1}A(SE) = E^{-1}BE = \begin{pmatrix} b_{11} & \delta b_{12} & 0 & & \cdots & 0 \\ 0 & b_{22} & \delta b_{23} & 0 & \cdots & 0 \\ \cdot & & \cdot & & & \cdot \\ \cdot & & & \cdot & & \cdot \\ \cdot & & & & \cdot & \delta b_{n-1,n} \\ 0 & \cdots & & & 0 & b_{nn} \end{pmatrix}$$

so that the off-diagonal elements can be made small by choosing δ small.

4   Special Classes of Matrices; Commutativity

4.1   Bilinear functional

Let U be a vector space (see **2.18**) and let β be a function on ordered pairs of vectors from U with values in **C** satisfying

(i) $\beta(cu_1 + du_2, v) = c\beta(u_1, v) + d\beta(u_2, v),$
(ii) $\beta(u, cv_1 + dv_2) = c\beta(u, v_1) + d\beta(u, v_2)$

for all vectors u_1, u_2, v_1, v_2, u, and v and all numbers c and d. Then β is called a *bilinear functional*. If statement (ii) is changed to

(ii)' $\beta(u, cv_1 + dv_2) = \bar{c}\beta(u, v_1) + \bar{d}\beta(u, v_2),$

then β is called a *conjugate bilinear functional*.

4.2   Examples

4.2.1   If U is the totality of n-vectors over **C**, i.e., $U = \mathbf{C}^n$ and A is any complex matrix, then

$$\beta(u, v) = \sum_{i,j=1}^{n} a_{ij} u_i \bar{v}_j,$$

where $u = (u_1, \cdots, u_n)$, $v = (v_1, \cdots, v_n)$, is a conjugate bilinear functional.

4.2.2   As indicated in the example **2.2.1**, $M_n(\mathbf{C})$ can be regarded as a vector space of n^2-tuples. The function $\beta(A,B) = \text{tr }(AB^*)$ is a conjugate bilinear functional. This follows immediately from the properties of the trace given in **2.8.1**.

4.3   Inner product

If U is a vector space and β is a conjugate bilinear functional which satisfies

(i) $\beta(u, v) = \overline{\beta(v, u)},$
(ii) $\beta(u, u) \geq 0$ with equality if and only if $u = 0$ (β is *positive definite*),

then β is called an *inner product* on U. The vector space U together with the functional β is called a *unitary space*. In case the underlying field is the real number field $\mathbf{R}$ then it will be assumed that β takes on just real values and condition (i) becomes simply $\beta(u, v) = \beta(v, u)$. In this case U together with β is called a *Euclidean space*. The notation (u, v) is commonly used for the inner product $\beta(u, v)$. The nonnegative number $(u, u)^{1/2}$ is called the *length* of u and is designated by $||u||$. The term *norm* of u is also used to designate $||u||$.

4.4 Example

In the vector space $\mathbf{C}^n$ of n-tuples of complex numbers,

$$(u, v) = \sum_{i=1}^{n} u_i \bar{v}_i, \qquad (u = (u_1, \cdots, u_n), v = (v_1, \cdots, v_n)),$$

is an inner product.

4.5 Orthogonality

If $u_1, \cdots, u_n$ are vectors in U and $(u_i, u_j) = 0$ for $i \neq j$, then $u_1, \cdots, u_n$ are said to be *pairwise orthogonal*. If, in addition $||u_i|| = 1$, $(i = 1, \cdots, n)$, then $u_1, \cdots, u_n$ is an *orthonormal set*. In general, if $||u|| = 1$, then u is a *unit* vector.

Some results of great importance for a unitary or Euclidean space follow.

4.5.1 (Gram-Schmidt orthonormalization process) If $x_1, \cdots, x_n$ are linearly independent, then there exists an orthonormal set $f_1, \cdots, f_n$ such that

$$\langle f_1, \cdots, f_k \rangle = \langle x_1, \cdots, x_k \rangle, \qquad (k = 1, \cdots, n). \tag{1}$$

Moreover, if $g_1, \cdots, g_n$ is any other orthonormal set satisfying $\langle g_1, \cdots, g_k \rangle = \langle x_1, \cdots, x_k \rangle$, $(k = 1, \cdots, n)$, then $g_i = c_i f_i$ where $|c_i| = 1$, $(i = 1, \cdots, n)$. This result, together with **3.11.2**, implies that any orthonormal set of k vectors in an n-dimensional space $(k < n)$ can be augmented to an orthonormal basis.

4.5.2 To construct the sequence $f_1, \cdots, f_n$ in **4.5.1**:

Step 1. Set $f_1 = x_1 ||x_1||^{-1}$.

Step 2. If $f_1, \cdots, f_k$ have been constructed, define $d_j = (x_{k+1}, f_j)$, $(j = 1, \cdots, k)$, and set $u_{k+1} = x_{k+1} - \sum_{j=1}^{k} d_j f_j$. Then

$$f_{k+1} = u_{k+1} ||u_{k+1}||^{-1}.$$

4.5.3 (Cauchy-Schwarz inequality) If u and ι

$$|(u, v)| \leq ||u||\,||v||$$

with equality if and only if u and v are linearly d

Proof: If $u = 0$, the inequality becomes an

$$w = v - \frac{(v, u)}{||u||^2}\, u.$$

Then

$$(w, u) = (v, u) - \frac{(v, u)}{||u||^2}\, (u, u) = 0$$

and

$$||w||^2 = (w, v) = ||v||^2 - \frac{(v, u)}{||u||^2}\, (u, v)$$

$$= ||v||^2 - \frac{|(u, v)|^2}{||u||^2}.$$

But

$$||w||^2 \geq 0$$

and therefore

$$||v||^2 \geq \frac{|(u, v)|^2}{||u||^2}$$

or, since $||u||$, $||v||$, and $|(u, v)|$ are nonnegative,

$$|(u, v)| \leq ||u||\,||v||.$$

The case of equality occurs, if and only if either $u = 0$ or $w = v - \frac{(v, u)}{||u||^2}\, u = 0$; i.e., if and only if u and v are linearly dependent.

4.5.4 (Triangle inequality) If u and v are any two vectors, then $||u + v|| \leq ||u|| + ||v||$ with equality only if one of the vectors is a non-negative multiple of the other.

Proof:
$$\begin{aligned}
||u + v||^2 &= (u + v, u + v)\\
&= ||u||^2 + (u, v) + (v, u) + ||v||^2\\
&= ||u||^2 + 2Re((u, v)) + ||v||^2\\
&\leq ||u||^2 + 2|(u, v)| + ||v||^2\\
&\leq ||u||^2 + 2||u||\,||v|| + ||v||^2, \quad \text{(by 4.5.3)},\\
&= (||u|| + ||v||)^2.
\end{aligned}$$

The equality occurs if and only if u and v are linearly dependent and $Re((u, v)) = |(u, v)|$; i.e., if and only if one of the vectors is a nonnegative multiple of the other.

4.5.5 (Pythagorean theorem) If $u_1, \cdots, u_n$ are orthonormal vectors and $x \in \langle u_1, \cdots, u_n \rangle$, then $x = \sum_{i=1}^{n} (x, u_i)u_i$ and $||x||^2 = \sum_{i=1}^{n} |(x, u_i)|^2$.

Example

As an example of the idea of inner product we will obtain two interesting formulas for the inner products of two skew-symmetric (2.7.3) and two symmetric (2.12.7) products of n-vectors, $(u_1 \wedge \cdots \wedge u_r, v_1 \wedge \cdots \wedge v_r)$ and $(u_1 \cdots u_r, v_1 \cdots v_r)$. We are, as usual, using the inner product described in 4.4 for all vectors considered.

Let $A \in M_{r,n}(\mathbf{C})$ and $B \in M_{r,n}(\mathbf{C})$ be the matrices whose rows are $u_1, \cdots, u_r$ and $v_1, \cdots, v_r$ respectively. Then, by 2.4.14,

$$
\begin{aligned}
(u_1 \wedge \cdots \wedge u_r, v_1 \wedge \cdots \wedge v_r) &= \sum_{\omega \in Q_{r,n}} d(A[1, \cdots, r|\omega]) \, \overline{d(B[1, \cdots, r|\omega])} \\
&= \sum_{\omega \in Q_{r,n}} d(A[1, \cdots, r|\omega]) \, d(B^*[\omega|1, \cdots, r]) \\
&= d(AB^*).
\end{aligned}
$$

But $AB^* \in M_r(\mathbf{C})$ and by matrix multiplication the (s,t) entry of AB^* is just (u_s, v_t). Hence

$$(u_1 \wedge \cdots \wedge u_r, v_1 \wedge \cdots \wedge v_r) = d((u_s, v_t)). \tag{1}$$

The situation for the symmetric product is analogous. For, by 2.11.7,
$(u_1 \cdots u_r, v_1 \cdots v_r)$

$$
\begin{aligned}
&= \sum_{\omega \in G_{r,n}} p(A[1, \cdots, r|\omega]) \, \overline{p(B[1, \cdots, r|\omega])}/\mu(\omega) \\
&= \sum_{\omega \in G_{r,n}} p(A[1, \cdots, r|\omega]) \, p(B^*[\omega|1, \cdots, r])/\mu(\omega) \\
&= p(AB^*) = p((u_s, v_t)), \qquad (s, t = 1, \cdots, r).
\end{aligned}
$$

Hence

$$(u_1 \cdots u_r, v_1 \cdots v_r) = p((u_s, v_t)). \tag{2}$$

From the formulas 4.6(1), 4.6(2), 2.7(8), and 2.12(1), we have that if $H \in M_n(\mathbf{C})$, then

$$(C_r(H)u_1 \wedge \cdots \wedge u_r, v_1 \wedge \cdots \wedge v_r) = d((Hu_i, v_j)) \tag{3}$$

and

$$(P_r(H)u_1 \cdots u_r, v_1 \cdots v_r) = p((Hu_i, v_j)). \tag{4}$$

4.7 Normal matrices

Let $A \in M_n(\mathbf{C})$. Then:

4.7.1 A is *normal* if $AA^* = A^*A$.

4.7.2 A is *hermitian* if $A^* = A$.

4.7.3 A is *unitary* if $A^* = A^{-1}$.

4.7.4 A is *skew-hermitian* if $A^* = -A$.

In case $A \in M_n(\mathbf{R})$, then the corresponding terms are as follows.

4.7.5 A is *real normal* if $AA^T = A^TA$.

4.7.6 A is *symmetric* if $A^T = A$.

4.7.7 A is *orthogonal* if $A^T = A^{-1}$.

4.7.8 A is *skew-symmetric* if $A^T = -A$.

4.7.9   If in what follows a result holds for the real as well as the complex case, we will not restate it for the real case. We will however sometimes consider matrices $A \in M_n(\mathbf{C})$ that are *complex symmetric*: $A^T = A$; *complex orthogonal*: $A^T = A^{-1}$; *complex skew-symmetric*: $A^T = -A$. When these arise we will carefully state exactly what is going on in order to avoid confusion with the real case.

Some elementary results concerning these classes of matrices are listed.

4.7.10   Hermitian or unitary or skew-hermitian matrices are normal.

4.7.11   If A_i are hermitian (skew-hermitian) and a_i are real numbers, $(i = 1, \cdots, m)$, then $\sum_{i=1}^{m} a_iA_i$ is hermitian (skew-hermitian).

4.7.12   If A is normal or hermitian or unitary, then so is A^p for any integer p (in case p is negative, then A is of course assumed nonsingular). If p is odd and A is skew-hermitian, then A^p is skew-hermitian.

4.7.13   If A and B are unitary, then AB is unitary. If A and B are normal and $AB = BA$, then AB is normal.

4.7.14   Let $(u, v) = \sum_{i=1}^{n} u_i\bar{v}_i$ be the inner product in the space $\mathbf{C}^n$ and suppose $A \in M_n(\mathbf{C})$. The fact that the equality $(Au, v) = (u, A^*v)$ holds for all u and v in $\mathbf{C}^n$ implies that A is

(i)   hermitian, if and only if $(Au, v) = (u, Av)$ for all u and v in $\mathbf{C}^n$ or (Au, u) is a real number for all u in $\mathbf{C}^n$;

(ii)   unitary, if and only if $||Au|| = ||u||$ for all $u \in \mathbf{C}^n$;

(iii)   skew-hermitian, if and only if $(Au, v) = -(u, Av)$ for all u and v in $\mathbf{C}^n$.

Similarly, if the inner product in the space $\mathbf{R}^n$ is taken to be $(u, v) = \sum_{i=1}^{n} u_iv_i$ and $A \in M_n(\mathbf{R})$, then $(Au, v) = (u, A^Tv)$ and it follows that A is

(iv)   symmetric, if and only if $(Au, v) = (u, Av)$ for all u and v in $\mathbf{R}^n$;

(v)   orthogonal, if and only if $||Au|| = ||u||$ for all $u \in \mathbf{R}^n$;

(vi)   skew-symmetric, if and only if $(Au, v) = -(u, Av)$ for all u and v in $\mathbf{R}^n$.

In what follows we will use the inner products for $\mathbf{C}^n$ and $\mathbf{R}^n$ described in **4.7.14** unless otherwise stated.

4.7.15   If $A \in M_n(\mathbf{C})$, then $A = H + iK$ where $H = \dfrac{A + A^*}{2}$ and $K = \dfrac{A - A^*}{2i}$ are both hermitian.

4.7.16   If $A \in M_n(\mathbf{R})$, then $A = S + T$ in which $S = \dfrac{A + A^T}{2}$ is symmetric and $T = \dfrac{A - A^T}{2}$ is skew-symmetric.

4.7.17   If $A \in M_n(\mathbf{C})$ and $u \in \mathbf{C}^n$ is a characteristic vector of A corresponding to the characteristic root r, then $Au = ru$ and hence $(Au, u) = (ru, u) = r(u, u)$. From this we can easily see the following results.

4.7.18   If A is hermitian or real symmetric, then any characteristic root of A is real [see **4.7.14**(i), (iv)].

4.7.19   If A is unitary or real orthogonal, then any characteristic root of A has absolute value 1 [see **4.7.14**(ii), (v)].

4.7.20   If A is skew-hermitian or real skew-symmetric, then any characteristic root of A has a zero real part [see **4.7.14**(iii), (vi)].

4.7.21   A normal matrix $A \in M_n(\mathbf{C})$ is hermitian if and only if the characteristic roots of A are real.

4.7.22   A normal matrix $A \in M_n(\mathbf{C})$ is skew-hermitian if and only if the characteristic roots of A have zero real part.

4.7.23   A normal matrix $A \in M_n(\mathbf{C})$ is unitary if and only if the characteristic roots of A have absolute value 1.

4.7.24   A matrix $A \in M_n(\mathbf{C})$ is unitary if and only if the rows (columns) of A form an orthonormal set of vectors in $\mathbf{C}^n$.

4.8   Examples

4.8.1   The criterion in **4.7.24** provides a method of constructing a unitary or orthogonal matrix with a preassigned unit vector as first row. For, given any unit vector $A_{(1)}$ in $\mathbf{C}^n$ or $\mathbf{R}^n$, we can use the Gram-Schmidt process (see **4.5.2**) to construct vectors $A_{(2)}, \cdots, A_{(n)}$ such that $A_{(1)}, \cdots, A_{(n)}$ form an orthonormal set.

4.8.2   Any n-square permutation matrix P is orthogonal. For, the rows of P are $e_1, \cdots, e_n$ (see **2.19**) in some order.

4.8.3 Let $P \in M_n(\mathbf{R})$ be the permutation matrix

$$
P = \begin{pmatrix}
0 & 0 & 0 & \cdots & 1 \\
1 & 0 & 0 & \cdots & 0 \\
0 & 1 & 0 & \cdots & 0 \\
\cdot & & & & \cdot \\
\cdot & & & & \cdot \\
\cdot & & & & \cdot \\
0 & & \cdots & 1 & 0
\end{pmatrix}. \tag{1}
$$

We compute the characteristic roots and characteristic vectors of P. Let e_j once again denote the n-tuple $(\delta_{1j}, \cdots, \delta_{nj})$. Then $Pe_1 = e_2$, $Pe_2 = e_3$, $\cdots$, $Pe_{n-1} = e_n$, $Pe_n = e_1$. Thus $P^n e_t = e_t$, $(t = 1, \cdots, n)$, and it follows that $P^n = I_n$. Hence the polynomial $\varphi(\lambda) = \lambda^n - 1$ has the property that $\varphi(P) = 0_{n,n}$ and therefore the minimal polynomial of P must divide $\varphi(\lambda)$ (see **3.28.1**). Suppose that $\sum_{t=0}^{n-1} c_t P^t = 0_{n,n}$ so that

$$
\sum_{t=0}^{n-1} c_t P^t = \begin{pmatrix}
c_0 & c_{n-1} & c_{n-2} & c_{n-3} & \cdots & c_1 \\
c_1 & c_0 & c_{n-1} & c_{n-2} & \cdots & c_2 \\
c_2 & c_1 & c_0 & c_{n-1} & \cdots & c_3 \\
\cdot & \cdot & \cdot & c_0 & & \cdot \\
\cdot & \cdot & \cdot & \cdot & \cdot & \cdot \\
\cdot & \cdot & \cdot & \cdot & & \cdot \\
c_{n-1} & c_{n-2} & c_{n-3} & c_{n-4} & \cdots & c_0
\end{pmatrix} = 0_{n,n}
$$

and thus $c_0 = \cdots = c_{n-1} = 0$. Hence no polynomial $\theta(\lambda)$ of degree $n - 1$ has the property that $\theta(P) = 0_{n,n}$. It follows that the minimum and characteristic polynomials of P are both $\varphi(\lambda) = \lambda^n - 1$ (P is nonderogatory as in **3.28.2**). Thus the characteristic roots of P are the nth roots of unity $r_k = \epsilon^k$, $(k = 0, \cdots, n - 1)$, where $\epsilon = e^{i2\pi/n}$. Next define

$$
u_t = \sum_{k=1}^{n} \epsilon^{kt} e_k, \qquad (t = 1, \cdots, n), \tag{2}
$$

so that

$$
\begin{aligned}
Pu_t &= \sum_{k=1}^{n} \epsilon^{kt} Pe_k \\
&= \epsilon^t e_2 + \epsilon^{2t} e_3 + \cdots + \epsilon^{(n-1)t} e_n + \epsilon^{nt} e_1 \\
&= \epsilon^{-t}(\epsilon^{(n+1)t} e_1 + \epsilon^{2t} e_2 + \cdots + \epsilon^{nt} e_n) \\
&= \epsilon^{-t} \sum_{k=1}^{n} \epsilon^{kt} e_k \\
&= \epsilon^{-t} u_t = \epsilon^{n-t} u_t.
\end{aligned} \tag{3}
$$

Moreover, for $s \neq t$

$$
(u_s, u_t) = \sum_{k=1}^{n} \epsilon^{(s-t)k} = \frac{\epsilon^{(s-t)} - \epsilon^{(s-t)(n+1)}}{1 - \epsilon^{(s-t)}} = 0, \tag{4}
$$

whereas

$$\|u_s\|^2 = (u_s, u_s) = n. \tag{5}$$

Hence combining the statements **4.8(3), 4.8(4), 4.8(5)** we can say that P has an orthonormal set of characteristic vectors $v_t = u_{n-t}/\sqrt{n}$ such that

$$Pv_t = \epsilon^t v_t, \qquad (t = 1, \cdots, n). \tag{6}$$

4.9   Circulant

As an interesting application of example **4.8.3**, we can analyze completely the structure of the characteristic roots and characteristic vectors for any matrix of the form

$$A = \begin{pmatrix} c_0 & c_{n-1} & c_{n-2} & \cdots & c_1 \\ c_1 & c_0 & c_{n-1} & \cdots & c_2 \\ \cdot & \cdot & c_0 & & \cdot \\ \cdot & \cdot & \cdot & \cdot & \cdot \\ \cdot & \cdot & & \cdot & \cdot \\ & & & & \cdot \\ c_{n-1} & c_{n-2} & c_{n-3} & \cdots & c_0 \end{pmatrix}.$$

It is clear that $A = \sum_{t=0}^{n-1} c_t P^t$ and hence using the notation of **4.8.3**

$$Av_k = \sum_{t=0}^{n-1} c_t P^t v_k$$

$$= \left(\sum_{t=0}^{n-1} c_t \epsilon^{kt} \right) v_k.$$

Thus if $\psi(\lambda)$ is the polynomial $c_0 + c_1\lambda + \cdots + c_{n-1}\lambda^{n-1}$, then $Av_k = \psi(\epsilon^k)v_k$ and A has characteristic vectors $v_1, \cdots, v_n$ corresponding respectively to the characteristic roots $\psi(\epsilon), \cdots, \psi(\epsilon^n)$. Any polynomial in the matrix P, $A = \sum_{t=0}^{n-1} c_t P^t$, is called a *circulant*.

4.10   Unitary similarity

4.10.1   If $A \in M_n(\mathbf{C})$ and $U \in M_n(\mathbf{C})$ and U is unitary, then $U^* = U^{-1}$ and $(U^*AU)_{ij} = (AU^{(j)}, U^{(i)})$, $(i, j = 1, \cdots, n)$. Suppose that A has an orthonormal set of characteristic vectors $u_1, \cdots, u_n$ corresponding respectively to characteristic roots $r_1, \cdots, r_n$. Let U be the matrix whose jth column is the n-tuple u_j, $(j = 1, \cdots, n)$. Then $U^*U = I_n$ so U is unitary and $(U^*AU)_{ij} = (AU^{(j)}, U^{(i)}) = (Au_j, u_i) = (r_j u_j, u_i) = r_j \delta_{ij}$. Hence $U^*AU = \text{diag}(r_1, \cdots, r_n)$. Thus a matrix is similar to a diagonal matrix

via a unitary matrix if and only if it has an orthonormal set of characteristic vectors (see also **3.12.3**).

4.10.2   (**Schur triangularization theorem**)   If $A \in M_n(\mathbf{C})$, then there exists
a unitary matrix $U \in M_n(\mathbf{C})$ such that $T = UAU^*$ is an upper triangular
matrix with the characteristic roots of A along the main diagonal. If
$A \in M_n(\mathbf{R})$ and A has real characteristic roots, then U may be chosen to
be a real orthogonal matrix. The matrix A is normal, if and only if T is
diagonal.

Proof: We use induction on n. Let x be a characteristic vector of A of
unit length, $Ax = \lambda_1 x$. Let V be any unitary matrix with x for its first
column (see **4.8.1**). Then $(V^*AV)^{(1)} = (V^*A)V^{(1)} = V^*Ax = \lambda_1 V^*x = \lambda_1 V^*V^{(1)} = \lambda_1(V^*V)^{(1)} = \lambda_1 e_1$ where e_1 denotes the n-tuple with 1 as its
first coordinate and 0 elsewhere. Thus

$$V^*AV = \begin{pmatrix} \lambda_1 & Y \\ \hline 0 & A_1 \end{pmatrix}$$

where $A_1 \in M_{n-1}(\mathbf{C})$ and $Y \in M_{1,n-1}(\mathbf{C})$. By the induction hypothesis,
there exists a unitary matrix W in $M_{n-1}(\mathbf{C})$ such that W^*A_1W is upper
triangular. Let $U = V(I_1 \dotplus W)$. Then U is unitary (see **4.7.13**) and U^*AU
is upper triangular. If $A \in M_n(\mathbf{R})$ and its characteristic roots are real,
then x can be chosen to be a real vector and therefore V and also U can be
chosen to be real orthogonal.

4.10.3   If $A \in M_n(\mathbf{C})$, then A is normal, if and only if there exists a unitary
matrix U such that $UAU^* = \operatorname{diag}(r_1, \cdots, r_n)$ where r_j, $(j = 1, \cdots, n)$,
are the characteristic roots of A. Thus hermitian, unitary, and skew her-
mitian matrices (all of which are normal) can be reduced to diagonal form
via a unitary matrix. It follows from **4.10.1** that A is normal if and only if
a set of characteristic vectors of A forms an orthonormal basis for $\mathbf{C}^n$.

4.10.4   Let $A \in M_n(\mathbf{R})$ be normal. Suppose $r_1, \cdots, r_k$ are the real characteristic roots of A and $c_t \pm id_t$, $(t = 1, \cdots, p)$, $2p + k = n$, are the complex characteristic roots of A (which occur in conjugate complex pairs
because the characteristic polynomial has real coefficients). Then there
exists a real orthogonal matrix $U \in M_n(\mathbf{R})$ such that

$$U^TAU = \operatorname{diag}(r_1, \cdots, r_k) \dotplus \sum_{t=1}^{p} \begin{pmatrix} c_t & -d_t \\ d_t & c_t \end{pmatrix}.$$

4.10.5   If $A \in M_n(\mathbf{R})$ is a real symmetric matrix, then there exists a real
orthogonal matrix $U \in M_n(\mathbf{R})$ such that $U^TAU = \operatorname{diag}(r_1, \cdots, r_n)$ where
r_j, $(j = 1, \cdots, n)$, are the characteristic roots of A. This result is of course
a special case of **4.10.4** since any real symmetric matrix is real normal and
has real characteristic roots (see **4.7.18**).

4.10.6 If $A \in M_n(\mathbf{C})$ and $A^T = A$, then there exists a unitary matrix $U \in M_n(\mathbf{C})$ such that $U^T A U = \text{diag}(r_1, \cdots, r_n)$. The numbers $r_1, \cdots, r_n$ satisfy $r_j \geq 0$, $r_j^2 = \lambda_j(A^*A)$, $(j = 1, \cdots, n)$.

4.10.7 If $A \in M_n(\mathbf{C})$ and $A^T = -A$, then there exists a unitary matrix U such that

$$U^*AU = 0_{n-2k, n-2k} \dotplus \sum_{t=1}^{k} \cdot \begin{pmatrix} 0 & a_t \\ -a_t & 0 \end{pmatrix}$$

where $a_t > 0$ and $a_t^2 = \lambda_t(A^*A)$, $(t = 1, \cdots, k)$, are the nonzero characteristic roots of A^*A.

4.10.8 Let A be an n-square permutation matrix (see **2.3**): $a_{i\sigma(i)} = 1$, $(i = 1, \cdots, n)$, $a_{ij} = 0$ otherwise. Suppose the permutation σ has cycle structure $[p_1, \cdots, p_m]$ (see **2.3**). Then there exists an n-square permutation matrix P such that $P^{-1}AP = P^T A P = \sum_{j=1}^{m} \cdot A_j$ in which A_j is p_j-square and of the form **4.8(1)** [if $p_j = 1$ then A_j is the 1-square matrix (1)]. In view of **4.8.3** and **2.15.10**, the characteristic roots of the permutation matrix A can be determined solely from the cycle structure of σ. More specifically, the characteristic roots of A are just the totality of roots of the m polynomials $\lambda^{p_t} - 1$, $(t = 1, \cdots, m)$.

4.11 Example

We compute the characteristic roots of the permutation matrix

$$A = \begin{pmatrix} 0 & 1 & 0 & 0 & 0 \\ 0 & 0 & 1 & 0 & 0 \\ 1 & 0 & 0 & 0 & 0 \\ 0 & 0 & 0 & 0 & 1 \\ 0 & 0 & 0 & 1 & 0 \end{pmatrix}.$$

The permutation σ for which $a_{i\sigma(i)} = 1$ is given by $\sigma(1) = 2$, $\sigma(2) = 3$, $\sigma(3) = 1$, $\sigma(4) = 5$, $\sigma(5) = 4$. The cycle structure of σ is $[3, 2]$ and from **4.10.8** there exists a permutation matrix P for which

$$P^{-1}AP = \begin{pmatrix} 0 & 0 & 1 & 0 & 0 \\ 1 & 0 & 0 & 0 & 0 \\ 0 & 1 & 0 & 0 & 0 \\ 0 & 0 & 0 & 0 & 1 \\ 0 & 0 & 0 & 1 & 0 \end{pmatrix}.$$

Moreover, the characteristic roots of A are $e^{i2\pi k/3}$, $(k = 0, 1, 2)$, and $e^{i2\pi k/2}$, $(k = 0, 1)$.

We next introduce an important class of hermitian matrices.

4.12 Positive definite matrices

If $A \in M_n(\mathbf{C})$ and A is hermitian, then A is called

(i) *positive (negative) definite*, if every characteristic root of A is positive (negative);

(ii) *positive (negative) semidefinite* or *nonnegative*, if every characteristic root of A is nonnegative (nonpositive).

4.12.1 (Hermitian form) If A is a hermitian matrix in $M_n(\mathbf{C})$, then the *hermitian form* associated with A is just the function $f(x) = \sum\limits_{i,j=1}^{n} a_{ij}x_i\bar{x}_j$ defined for all $x \in \mathbf{C}^n$. Then A is positive definite if and only if $f(x) > 0$ for all $x \neq 0_n$. To see this let $u = (\bar{u}_1, \cdots, \bar{u}_n)$ where $(u_1, \cdots, u_n)$ is a characteristic vector of A corresponding to the characteristic root r. Then

$$f(u) = \sum_{i,j=1}^{n} a_{ij}\bar{u}_i u_j = \sum_{i=1}^{n} \left(\sum_{j=1}^{n} a_{ij}u_j \right) \bar{u}_i$$

$$= \sum_{i=1}^{n} ru_i\bar{u}_i = r \sum_{i=1}^{n} |u_i|^2.$$

If $f(u) > 0$, then $r > 0$ and it follows that A is positive definite. Suppose conversely that A is positive definite. According to **4.10.3** there exists a unitary matrix U such that

$$A = U^* \operatorname{diag}(r_1, \cdots, r_n)U, \qquad (r_i > 0; i = 1, \cdots, n).$$

If $x = (x_1, \cdots, x_n)$ is any vector in $\mathbf{C}^n$, let $y = U\bar{x}$ and observe that

$$f(x) = \sum_{i,j=1}^{n} a_{ij}x_i\bar{x}_j = \sum_{i,j=1}^{n} \left(\sum_{s=1}^{n} \bar{u}_{si}r_s u_{sj} \right) x_i\bar{x}_j$$

$$= \sum_{s=1}^{n} r_s \sum_{i=1}^{n} x_i \left(\sum_{j=1}^{n} u_{sj}\bar{x}_j \right) \bar{u}_{si}$$

$$= \sum_{s=1}^{n} r_s \left(\sum_{i=1}^{n} \bar{u}_{si}x_i \right) y_s = \sum_{s=1}^{n} r_s\bar{y}_s y_s = \sum_{s=1}^{n} r_s|y_s|^2.$$

Since $x = 0_n$ if and only if $y = 0_n$, it follows that $f(x) > 0$ for all $x \neq 0_n$.

4.12.2 If $A \in M_n(\mathbf{C})$, then A^*A and AA^* are positive semidefinite. Moreover, $\rho(A^*A) = \rho(AA^*) = \rho(A)$ (see **2.17.2**) and hence A^*A and AA^* are positive definite if and only if A is nonsingular. The characteristic roots of A^*A are all nonnegative and the numbers

$$\sqrt{\lambda_j(A^*A)}, \quad (j = 1, \cdots, n),$$

are called the *singular values* of A. Note that $(A^*Au, u) = (Au, Au) = \|Au\|^2 \geq 0$.

4.12.3 For a normal matrix A there is a close relationship between the singular values of A and the characteristic roots of A. If $A \in M_n(\mathbf{C})$ is normal, then the singular values of A are the characteristic roots of A in absolute value:

$$\sqrt{\lambda_j(A^*A)} = |\lambda_j(A)|, \qquad (j = 1, \cdots, n).$$

4.12.4 If $A \in M_n(\mathbf{C})$ is hermitian and $\rho(A) = p$, then A is positive semidefinite, if and only if there exists a permutation matrix $Q \in M_n(\mathbf{C})$ such that

$$d(Q^T A Q[1, \cdots, k | 1, \cdots, k]) > 0, \qquad (k = 1, \cdots, p). \qquad (1)$$

If $p = n$, then **4.12(1)** holds for $k = 1, \cdots, n$, if and only if A is positive definite.

4.12.5 A hermitian matrix is positive definite (semidefinite), if and only if every principal submatrix is positive definite (nonnegative) hermitian.

4.12.6 The Kronecker product, compound matrix, and induced matrix inherit certain properties from the component matrices. If $A \in M_n(\mathbf{C})$ and $x_1, \cdots, x_k$ are characteristic vectors of A corresponding respectively to the characteristic roots $r_1, \cdots, r_k$, then the skew-symmetric product

$$x_1 \wedge \cdots \wedge x_k$$

(see **2.7.3**) is a characteristic vector of $C_k(A)$ corresponding to the product $r_1 \cdots r_k$; the symmetric product $x_1 \cdots x_k$ (see **2.12.7**) is a characteristic vector of $P_k(A)$ corresponding to $r_1 \cdots r_k$. (Of course if the x_j are linearly dependent then $x_1 \wedge \cdots \wedge x_k = 0$; hence in the case of $C_k(A)$ we assume the x_j are linearly independent.)

4.12.7 If A and B are normal matrices in $M_n(\mathbf{C})$ and $M_m(\mathbf{C})$ respectively, then

(i) $A \otimes B$ is normal in $M_{mn}(\mathbf{C})$;

(ii) if $1 \leq r \leq n$, then $C_r(A)$ is a normal $\binom{n}{r}$-square matrix over $\mathbf{C}$ and $P_r(A)$ is a normal $\binom{n+r-1}{r}$-square matrix over $\mathbf{C}$.

Moreover, if A is any one of hermitian, positive definite hermitian, nonnegative hermitian, skew-hermitian (r odd), unitary, then so are $C_r(A)$ and $P_r(A)$. If A and B both have any of the preceding properties (with the exception of the skew-hermitian property), then so does the mn-square matrix $A \otimes B$.

4.13 Example

Let β be a conjugate bilinear functional defined on ordered pairs of vectors from $\mathbf{C}^n$. Let $u_1, \cdots, u_n$ be a basis for $\mathbf{C}^n$ and define an associated matrix $A \in M_n(\mathbf{C})$ by $a_{ij} = \beta(u_i, u_j)$. Then we can easily prove that β is positive definite if and only if A is positive definite hermitian. This justifies the use of the term "positive definite" in both cases. For if $x = (x_1, \cdots, x_n)$ is any vector in $\mathbf{C}^n$ and $u = \sum\limits_{i=1}^{n} x_i u_i$, then

$$\beta(u, u) = \beta \left(\sum_{i=1}^{n} x_i u_i, \sum_{i=1}^{n} x_i u_i \right) = \sum_{i,j=1}^{n} \beta(u_i, u_j) x_i \bar{x}_j = \sum_{i,j=1}^{n} a_{ij} x_i \bar{x}_j.$$

According to 4.12.1, $\sum\limits_{i,j=1}^{n} a_{ij} x_i \bar{x}_j > 0$ for $x \neq 0_n$ if and only if A is positive definite hermitian.

4.14 Functions of normal matrices

If $A \in M_n(\mathbf{C})$ is normal, then let U be a unitary matrix such that $U^* A U = \text{diag}\,(r_1, \cdots, r_n)$ where r_j, $(j = 1, \cdots, n)$, are the characteristic roots of A. If $f(z)$ is a function defined for $z = r_j$, $(j = 1, \cdots, n)$, then set

$$f(A) = U \,\text{diag}\,(f(r_1), \cdots, f(r_n)) U^*. \tag{1}$$

In case $f(z) = a_k z^k + a_{k-1} z^{k-1} + \cdots + a_1 z + a_0$,

$$\begin{aligned}
f(A) &= U \,\text{diag}\,(f(r_1), \cdots, f(r_n)) U^* \\
&= U \left(\sum_{t=0}^{k} a_t \,\text{diag}\,(r_1^t, \cdots, r_n^t) \right) U^* \\
&= \sum_{t=0}^{k} a_t U (\text{diag}\,(r_1, \cdots, r_n))^t U^* \\
&= \sum_{t=0}^{k} a_t (U \,\text{diag}\,(r_1, \cdots, r_n) U^*)^t \\
&= \sum_{t=0}^{k} a_t A^t.
\end{aligned} \tag{2}$$

Hence the definition 4.14(1) agrees with 2.13.1. Note that, by 4.14(1), $f(A)$ has $f(r_1), \cdots, f(r_n)$ as characteristic roots. It is also the case that when A is normal A^* is a polynomial in A.

4.14.1 If $A \in M_n(\mathbf{C})$ is hermitian positive semidefinite and $f(z) = z^{1/m}$, $(m > 0)$, then $f(A)$ satisfies $(f(A))^m = A$. If $f(z) = |z^{1/m}|$, then $f(A) = B$ is a uniquely determined positive semidefinite hermitian matrix and satisfies $B^m = A$. The matrix B is sometimes written $A^{1/m}$. Of particular interest is the *square root* of A, $A^{1/2}$.

4.15 Examples

4.15.1 If J is the matrix in $M_n(\mathbf{R})$ all of whose entries are 1 and $p > 0$, $p + nq > 0$, then the matrix $A = p I_n + q J$ is positive definite. In fact, the matrix A is a circulant (see 4.9) in which

$$c_0 = p + q, \qquad (c_1 = c_2 = \cdots = c_{n-1} = q):$$
$$A = (p + q)I_n + qP + qP^2 + \cdots + qP^{n-1}.$$

The characteristic roots of A are the numbers (see also **2.15.3**)

$$r_k = (p + q) + q\epsilon^k + q\epsilon^{2k} + \cdots + q\epsilon^{(n-1)k}$$
$$= p \text{ for } k = 1, \cdots, n - 1, \text{ and } r_n = p + nq.$$

(Recall that $\epsilon = e^{i2\pi/n}$.) The orthonormal characteristic vectors of A are the n-tuples $v_{n-k} = (\epsilon^k, \epsilon^{2k}, \epsilon^{3k}, \cdots, \epsilon^{nk})/\sqrt{n}$ where

$$Av_t = r_t v_t, \quad (t = 1, \cdots, n).$$

If U is the unitary matrix whose kth column is v_k, then

$$U^*AU = \operatorname{diag}(r_1, \cdots, r_n).$$

Then

$$
\begin{aligned}
A^{1/2} &= U \operatorname{diag}(r_1^{1/2}, \cdots, r_n^{1/2})U^* \\
&= U \operatorname{diag}(p^{1/2}, \cdots, p^{1/2}, (p + nq)^{1/2})U^* \\
&= U[p^{1/2}I_n + ((p + nq)^{1/2} - p^{1/2}) \operatorname{diag}(0, \cdots, 0, 1)]U^* \\
&= p^{1/2}I_n + ((p + nq)^{1/2} - p^{1/2})U \operatorname{diag}(0, \cdots, 0, 1)U^*.
\end{aligned}
$$

The matrix $U \operatorname{diag}(0, \cdots, 0, 1)U^*$ has as its (s,t) entry $u_{sn}\overline{u}_{nt} = n^{-1/2}n^{-1/2} = n^{-1}$. Hence $A^{1/2} = p^{1/2}I_n + [((p + nq)^{1/2} - p^{1/2})/n]J$.

4.15.2 From **2.4.10** a (v,k,λ)-matrix satisfies $AA^T = (k - \lambda)I_v + \lambda J$ where $J \in M_v(\mathbf{I})$ has every entry equal to 1. It was also proved in **2.4.10** that $A^{-1} = (k - \lambda)^{-1}(A^T - \lambda k^{-1}J)$, $\lambda = k(k - 1)(v - 1)^{-1}$, and $JA = kJ$. Thus

$$
\begin{aligned}
I_v = A^{-1}A &= (k - \lambda)^{-1}(A^TA - \lambda k^{-1}JA) \\
&= (k - \lambda)^{-1}(A^TA - \lambda J), \\
A^TA &= (k - \lambda)I_v + \lambda J = AA^T
\end{aligned}
$$

and it follows that A is normal.

4.16 Exponential of a matrix

If $A \in M_n(\mathbf{C})$ is normal, then $f(z) = \exp(z)$ is defined for all complex numbers and hence in particular for the characteristic roots of A. Thus $f(A) = \exp(A)$ is well defined according to **4.14**. Another definition of $\exp(A)$ can be made for an arbitrary $A \in M_n(\mathbf{C})$ as follows: it is a fact that $\sum_{k=0}^{\infty} (A^k)_{ij}/k!$ converges for any A and all (i,j). Then $\exp(A)$ is defined to be $\sum_{k=0}^{\infty} A^k/k!$. It turns out that in case A is normal, the two definitions are the same. A rather surprising formula is

$$d(\exp(A)) = \exp(\operatorname{tr}(A)). \tag{1}$$

4.17   Functions of an arbitrary matrix

Let $p(\lambda) = (\lambda - r)^k$ and let A be the k-square hypercompanion matrix of $p(\lambda)$ (see **3.29.1**). If $f(\lambda)$ is a polynomial, then $f(A)$ is defined by

$$f(A) = \begin{pmatrix} f(r) & \dfrac{f'(r)}{1!} & \dfrac{f''(r)}{2!} & \cdots & \dfrac{f^{(k-1)}(r)}{(k-1)!} \\ & & & & \\ & & & \dfrac{f''(r)}{2!} \\ & 0 & & \dfrac{f'(r)}{1!} \\ & & & f(r) \end{pmatrix} \in M_k(\mathbf{C}), \quad (1)$$

where $f^{(t)}(r)$ is the tth derivative of $f(\lambda)$ evaluated at r $(f^{(0)}(\lambda) = f(\lambda))$. If B is any matrix in $M_n(\mathbf{C})$, then B is similar over $\mathbf{C}$ to the direct sum of the hypercompanion matrices of all the elementary divisors of $\lambda I_n - B$ (see **3.29.4**):

$$B = S^{-1} \left(\sum_{i=1}^{m} \dot{}\, A_i \right) S.$$

Then $f(B)$ is defined by

$$f(B) = S^{-1} \left(\sum_{i=1}^{m} \dot{}\, f(A_i) \right) S. \quad (2)$$

More generally, if $f(\lambda)$ is any function for which $f^{(t)}(r)$, $(t = 0, \cdots, e(r) - 1)$, is defined for each characteristic root r of B where $(\lambda - r)^{e(r)}$ is the elementary divisor of highest degree involving $\lambda - r$, then the formulas **4.17(1),(2)** define $f(B)$. In the case of a normal matrix this definition coincides with those in **4.14** and **4.16**.

4.18   Example

Suppose $B \in M_n(\mathbf{C})$ and $f(\lambda)$ is a polynomial in $\mathbf{C}[\lambda]$ such that $f(0) = 0$ if 0 is a characteristic root of B. Then there exists a polynomial $q(\lambda) \in \mathbf{C}[\lambda]$ for which $q(B) = f(B)$ and $q(0) = 0$. For, in the notation of **4.17**, it suffices to choose $q(\lambda)$ so that

$$q^{(t)}(r) = f^{(t)}(r), \qquad (t = 0, \cdots, e(r) - 1), \quad (1)$$

for each characteristic root r of B. Suppose first that $d(B) \neq 0$ and define $p(\lambda) = (-1)^n \psi(\lambda) f(0)/d(B)$ where $\psi(\lambda)$ is the characteristic polynomial of B. Then set $q(\lambda) = f(\lambda) - p(\lambda)$ and observe by direct verification of 4.18(1) or use of 3.28.2 that

$$q(A) = f(A) \quad \text{and} \quad q(0) = f(0) - (1)^n \psi(0) f(0) / d(B) = 0.$$

If $d(B) = 0$, i.e., B has a zero characteristic root, then simply set $q(\lambda) = f(\lambda)$ and the condition $f(0) = 0$ automatically ensures that $q(0) = 0$.

4.19   Representation of a matrix as a function of other matrices

The exponential and other functions make their appearance in the following list of representation theorems.

4.19.1   The matrix $U \in M_n(\mathbf{C})$ is unitary, if and only if $U = \exp (iH)$ where $H \in M_n(\mathbf{C})$ is hermitian. If $U \in M_n(\mathbf{R})$ and $d(U) = 1$, then U is real orthogonal if and only if $U = \exp (K)$ where $K \in M_n(\mathbf{R})$ is skew-symmetric.

4.19.2   (Cayley parameterization)   If $U \in M_n(\mathbf{C})$ is unitary and does not have -1 as a characteristic root, then

$$U = (I_n + iH)(I_n - iH)^{-1}$$

where $H \in M_n(\mathbf{C})$ is hermitian and uniquely determined by the formula

$$H = i(I_n - U)(I_n + U)^{-1}.$$

4.19.3   If $U \in M_n(\mathbf{C})$ is unitary and $U^T = U$, then there exists $K \in M_n(\mathbf{R})$, $K^T = K$; i.e., K is a real symmetric matrix, such that $U = \exp (iK)$.

4.19.4   (Polar factorization)   If $A \in M_n(\mathbf{C})$, then there exist unique positive semidefinite hermitian matrices H and K and unitary matrices U and V all in $M_n(\mathbf{C})$ such that $A = UH = KV$. Moreover, $H = (A^*A)^{1/2}$, $K = (AA^*)^{1/2}$. The matrices U and V are uniquely determined if and only if A is nonsingular. A is normal if and only if $UH = HU = KV = VK = A$. If $A \in M_n(\mathbf{R})$, then all of U, V, H, and K may be taken to be real matrices.

Proof: Let $r_1^2, \cdots, r_n^2$ be the characteristic roots of the positive semidefinite matrix A^*A (see 4.12.2),

$$r_j > 0, \quad (j = 1, \cdots, k), \text{ and } r_j = 0, \ (j = k + 1, \cdots, n).$$

Let $x_1, \cdots, x_n$ be the corresponding orthonormal characteristic vectors of A^*A (see 4.10.3). Then, for $1 < i, j \leq k$,

$$\left(\frac{Ax_i}{r_i}, \frac{Ax_j}{r_j} \right) = \frac{(A^*Ax_i, x_j)}{r_i r_j} = \frac{\delta_{ij} r_i^2}{r_i r_j}$$

and thus the vectors $z_j = Ax_j / r_j$, $(j = 1, \cdots, k)$, are orthonormal. Let X

and Z be unitary matrices in $M_n(\mathbf{C})$ such that $X^{(j)} = x_j$, $(j = 1, \cdots, n)$, and $Z^{(j)} = z_j$, $(j = 1, \cdots, k)$. We have

$$AX^{(j)} = r_j Z^{(j)}, \qquad (j = 1, \cdots, n),$$

or, with $R = \mathrm{diag}\ (r_1, \cdots, r_n)$,

$$AX = ZR.$$

Now, let $U = ZX^*$ and $H = XRX^*$. Clearly U is unitary and H is positive semidefinite hermitian. Furthermore,

$$UH = ZX^*XRX^* = ZRX^* = A.$$

Applying the above result to A^* we obtain $A = KV$.

4.19.5 If $U \in M_n(\mathbf{C})$ and $U^T = U^{-1}$, then $U = V \exp\ (iH)$ in which $V \in M_n(\mathbf{R})$ is real orthogonal and $H \in M_n(\mathbf{R})$ is real skew-symmetric.

4.19.6 If $A \in M_n(\mathbf{C})$ is nonsingular, then there exist H, K, U, and V in $M_n(\mathbf{C})$ such that $H^T = H, K^T = K, U^T = U^{-1}, V^T = V^{-1}, A = UH = KV$, $A^T A = H^2, AA^T = K^2$. Moreover, $AA^T = A^T A$ if and only if $UH = HU = KV = VK = A$.

4.19.7 If $A \in M_n(\mathbf{C})$ is nonsingular, then A is *circular* if $\overline{A} = A^{-1}$. The matrix $A = B + iD$, (B and D real n-square matrices), is circular, if and only if $BD = DB$, $B^2 + D^2 = I_n$. Moreover, A is circular if and only if $A = \exp\ (iK)$ where $K \in M_n(\mathbf{R})$.

4.19.8 If $A \in M_n(\mathbf{C})$, then $A = GH$ where $G \in M_n(\mathbf{C})$ is circular and $H \in M_n(\mathbf{R})$. Moreover, G can be chosen so that $\lambda_j(G) + \overline{\lambda_j(G)} > 0$, $(j = 1, \cdots, n)$, in which case the factors G and H are uniquely determined.

4.19.9 Let $E \in M_2(\mathbf{R})$ be the matrix $E = \begin{pmatrix} 0 & 1 \\ -1 & 0 \end{pmatrix}$. Set

$$J = E \otimes I_n \in M_{2n}(\mathbf{R}).$$

Then $J^2 = E^2 \otimes I_n = -I_2 \otimes I_n = -I_{2n}$. A matrix $A \in M_{2n}(\mathbf{R})$ is called *Hamiltonian*, if $(JA)^T = JA$. A matrix $A \in M_{2n}(\mathbf{R})$ is called *symplectic* if $A^T J A = J$.

4.20 Examples

4.20.1 As an application of much of the above material we can show quite easily that if $A \in M_n(\mathbf{C})$ and all r-square principal subdeterminants of A are 1, for some fixed r, $(1 \le r < n)$, while all the remaining r-square subdeterminants of A are 0, then $A = cI_n$ where $c^r = 1$. The conditions on A amount to saying that $C_r(A) = I_{\binom{n}{r}}$. By the Sylvester-Franke theorem [**2.7(6)**], A must be nonsingular and hence, by **4.19.4**, A can be written uniquely as $A = UH$ in which U is unitary and H is positive

definite hermitian. Thus $I_{\binom{n}{r}} = C_r(A) = C_r(UH) = C_r(U)C_r(H)$ and, by
4.12.7, $C_r(U)$ is unitary and $C_r(H)$ is positive definite hermitian. But
$(C_r(U))^{-1} = C_r(H)$ and hence $C_r(U)$ and $C_r(H)$ are both unitary and
positive definite hermitian. It follows that the characteristic roots of both
$C_r(U)$ and $C_r(H)$ must be positive and on the unit circle. Thus every
characteristic root of $C_r(U)$ and $C_r(H)$ has value 1. Let $\theta_1, \cdots, \theta_n$ and
$h_1, \cdots, h_n$ be characteristic roots of U and H respectively. By **2.15.12,**
all $\binom{n}{r}$ products $\theta_{i_1} \cdots \theta_{i_r}$ as well as the $\binom{n}{r}$ products $h_{i_1} \cdots h_{i_r}$,
$1 \le i_1 < \cdots < i_r \le n$ have value 1. Since $r < n$, it follows that
$\theta_1 = \cdots = \theta_n$ and $h_1 = \cdots = h_n$. The h_i are positive and hence

$$h_1 = \cdots = h_n = 1;$$

also $\theta_1 = \cdots = \theta_n = c$, $(c^r = 1)$. By **4.10.3,** there exists an n-square
unitary matrix V for which $VHV^* = I_n$ and it follows that $H = I_n$.
Similarly, from **4.10.3,** $U = cI_n$ and hence $A = UH = cI_n$, $(c^r = 1)$, as
announced above.

4.20.2 We show that $A \in M_{2n}(\mathbf{R})$ is orthogonal and symplectic if and
only if $A = I_2 \otimes U + E \otimes V$ in which U and V are in $M_n(\mathbf{R})$, $U + iV$ is
unitary and E is the matrix in **4.19.9.** For $A^T = A^{-1}$ and hence $JA = AJ$.
If we write

$$A = \begin{pmatrix} A_{11} & A_{12} \\ A_{21} & A_{22} \end{pmatrix}$$

in which $A_{ij} \in M_n(\mathbf{R})$, $(i, j = 1, 2)$, then $AJ = JA$ if and only if $A_{11} =$
$A_{22} = U$ and $A_{12} = -A_{21} = V$. Hence A has the form

$$A = I_2 \otimes U + E \otimes V.$$

Then

$$\begin{aligned} AA^T &= (I_2 \otimes U + E \otimes V)(I_2 \otimes U^T + E^T \otimes V^T) \\ &= I_2 \otimes (UU^T + VV^T) + E \otimes (VU^T - UV^T). \end{aligned}$$

It follows that $AA^T = I_{2n}$ if and only if $UU^T + VV^T = I_n$ and
$VU^T - UV^T = 0_{n,n}$. But these are precisely the necessary and sufficient
conditions for $U + iV$ to be unitary as can be readily checked by multiplying out $(U + iV)(U + iV)^* = I_n$.

4.20.3 As an application of Schur's theorem on complex symmetric
matrices (see **4.10.6**) we can prove easily that if $A \in M_{2n}(\mathbf{R})$ is Hamiltonian
and symmetric, then there exists an orthogonal symplectic $T \in M_{2n}(\mathbf{R})$
such that $T^{-1}AT = F \otimes D$ in which $F = \begin{pmatrix} 1 & 0 \\ 0 & -1 \end{pmatrix}$ and $D \in M_n(\mathbf{R})$ is a
diagonal matrix. It follows by partitioning A, as was done in **4.20.2,** that
$A = F \otimes M + E \otimes L$ in which M and L are real symmetric n-square
matrices. Let H be the complex n-square symmetric matrix $M + iL$. Then,
by **4.10.6,** there exists a unitary matrix $S = U + iV$, (U and V n-square

real), such that $S^T(M + iL)S = D$ in which D is a real diagonal matrix with the singular values of $M + iL$ on the main diagonal. The matrix $T = I_2 \otimes U + E \otimes V$ is orthogonal symplectic and from the equation $(M + iL)(U + iV) = (U + iV)D$ we conclude immediately that

$$AT = \begin{pmatrix} UD & -VD \\ -VD & -UD \end{pmatrix} = T(F \otimes D).$$

4.21   Simultaneous reduction of commuting matrices

We go on now to three results which are extensions of **4.10.2, 4.10.3,** and **4.10.4** respectively.

4.21.1   If $A_i \in M_n(\mathbf{C})$, $(i = 1, \cdots, m)$, pairwise commute [i.e., $A_i A_j = A_j A_i$, $(i, j = 1, \cdots, m)$], then there exists a unitary matrix $U \in M_n(\mathbf{C})$ such that $T_i = U^* A_i U$ is an upper triangular matrix for $i = 1, \cdots, m$.

4.21.2   Using the same notation, it follows that if the matrices A_i in **4.21.1** are normal, then the matrices T_i are diagonal for $i = 1, \cdots, m$.

4.21.3   If $A_j \in M_n(\mathbf{R})$, $(j = 1, \cdots, m)$, are pairwise commutative real normal matrices, then there exists a real orthogonal matrix $U \in M_n(\mathbf{R})$ such that

$$U^T A_j U = \mathrm{diag}\,(r_{j1}, \cdots, r_{jk_i}) \dotplus \sum_{t=1}^{p_i} \cdot \begin{pmatrix} c_{jt} & -d_{jt} \\ d_{jt} & c_{jt} \end{pmatrix}$$

in which A_j has real characteristic roots $r_{j1}, \cdots, r_{jk_i}$ and (possibly) complex roots

$$c_{j1} \pm i d_{j1}, \cdots, c_{jp_i} \pm i d_{jp_i}, \qquad (j = 1, \cdots, m).$$

Note that d_{jt} may be 0 for some j and t since the matrices A_j are not in general likely to have the same number of real characteristic roots simply because they commute.

4.22   Commutativity

4.22.1   If $A_i \in M_n(\mathbf{C})$, $(i = 1, \cdots, m)$, are pairwise commutative hermitian matrices, then there exists a hermitian matrix $H \in M_n(\mathbf{C})$ and m polynomials $p_i(\lambda)$, $(i = 1, \cdots, m)$, with real coefficients such that $A_i = p_i(H)$, $(i = 1, \cdots, m)$.

4.22.2   If A and B are in $M_n(\mathbf{C})$, both commute with $AB - BA$, and A has k distinct characteristic roots, then there exists a polynomial $p(\lambda) \in \mathbf{C}[\lambda]$ such that $B - p(A)$ has at least k linearly independent characteristic vectors corresponding to the characteristic root 0 [i.e., in the null space of $B - p(A)$]. It follows that if $k = n$, then B is a polynomial in A.

4.22.3 If A and B are commuting matrices in $M_n(\mathbf{C})$ and A is non-derogatory, then B is a polynomial in A.

4.22.4 If A and B are in $M_n(\mathbf{C})$ and B commutes with every $X \in M_n(\mathbf{C})$ that commutes with A, then B is a polynomial in A.

4.23 Example

Suppose that A and B are given by

$$A = \begin{pmatrix} 0 & 0 & 0 \\ 2 & 0 & 0 \\ 0 & 0 & 0 \end{pmatrix}, \qquad B = \begin{pmatrix} 0 & 0 & 0 \\ 0 & 0 & 3 \\ 0 & 0 & 4 \end{pmatrix}.$$

Then $AB = 0_{3,3} = BA$ but clearly neither matrix is a polynomial in the other. Moreover, they cannot be expressed as polynomials in some third matrix C.

4.24 Quasi-commutativity

4.24.1 Let $A_1, \cdots, A_m$ be matrices in $M_n(\mathbf{C})$ and let $p(x_1, \cdots, x_m)$ be a polynomial over $\mathbf{C}$ in the noncommuting variables $x_1, \cdots, x_m$: e.g., $m = 2$ and $p(x_1, x_2) = x_1 x_2 - x_2 x_1 + x_1$. Then the matrices $A_1, \cdots, A_m$ are called *quasi-commutative*, if $p(A_1, \cdots, A_m)(A_i A_j - A_j A_i)$ is nilpotent (see **1.5.3**) for any such polynomial $p(x_1, \cdots, x_m)$ and all i, j.

4.24.2 In the notation of **4.24.1**, the following conditions are equivalent to quasi-commutativity:

(i)   There exists a unitary $U \in M_n(\mathbf{C})$ such that $U^* A_i U$ is upper triangular for $i = 1, \cdots, m$.

(ii) There is an ordering of the characteristic roots of $A_i : \lambda_{ik}$, $(k = 1, \cdots, n; i = 1, \cdots, m)$, such that the following holds. Let $p(x_1, \cdots, x_m)$ and $q(x_1, \cdots, x_m)$ be any two polynomials over $\mathbf{C}$ in the noncommuting variables $x_1, \cdots, x_m$ such that $B = q(A_1, \cdots, A_m)$ is nonsingular. Then the characteristic roots of $p(A_1, \cdots, A_m)B^{-1}$ are $p(\lambda_{1k}, \cdots, \lambda_{mk})/q(\lambda_{1k}, \cdots, \lambda_{mk})$, $(k = 1, \cdots, n)$.

4.24.3 Let $A_i \in M_n(\mathbf{C})$, $(i = 1, \cdots, m)$. The following statements are equivalent:

(i) $A_i A_j = A_j A_i$ for all i and j and for each A_i there exists a matrix $P_i \in M_n(\mathbf{C})$ such that $P_i^{-1} A_i P_i$ is a diagonal matrix, $(i = 1, \cdots, m)$.

(ii) There exists a matrix $P \in M_n(\mathbf{C})$ such that $P^{-1} A_i P$ is a diagonal matrix, $(i = 1, \cdots, m)$.

(iii) There exist a matrix A and m polynomials $f_i(\lambda) \in \mathbf{C}[\lambda]$ such that $A_i = f_i(A)$ and A is similar to a diagonal matrix over $\mathbf{C}$.

4.25 Example

The matrices $A_1 = \begin{pmatrix} 0 & 1 \\ 0 & 0 \end{pmatrix}$ and $A_2 = \begin{pmatrix} 3 & 1 \\ 0 & 2 \end{pmatrix}$ are quasi-commutative but not commutative.

4.26 Property L

Let A and B be in $M_n(\mathbf{C})$. Suppose that there exists an ordering of the characteristic roots of A and B, $r_1, \cdots, r_n$ and $s_1, \cdots, s_n$, respectively, such that for all complex numbers c and h the matrix $cA + hB$ has as characteristic roots $cr_i + hs_i$, $(i = 1, \cdots, n)$. Then A and B are said to have *property L*. The totality of matrices $cA + hB$ is called the *pencil generated* by A and B.

4.26.1 If A and B are in $M_2(\mathbf{C})$ and have property L, then A and B are quasi-commutative.

4.26.2 If A and B are in $M_n(\mathbf{C})$ and have property L $(n > 2)$, then either every matrix in the pencil generated by A and B has a characteristic root of algebraic multiplicity at least 2, or at most $n(n-1)/2$ of the matrices in the pencil have this property.

4.26.3 If every matrix in the pencil generated by A and B is similar over $\mathbf{C}$ to a diagonal matrix, then $AB = BA$.

4.26.4 If $AB = BA$, then the pencil generated by A and B

(i)   consists only of matrices similar to a diagonal matrix, or
(ii) has no matrices in it similar to a diagonal matrix, or
(iii) has exactly one matrix in it similar to a diagonal matrix.

For 2-square matrices the alternative (ii) is not possible.

4.27 Examples

4.27.1 If

$$A = \begin{pmatrix} 0 & 1 & 0 \\ 0 & 0 & 1 \\ 0 & 0 & 0 \end{pmatrix} \quad \text{and} \quad B = \begin{pmatrix} 1 & 1 & 2 \\ 1 & 1 & 2 \\ -1 & -1 & -2 \end{pmatrix},$$

then A and B both have 0 as a characteristic root of algebraic multiplicity 3, whereas $A + B$ is nonsingular. Thus property L fails for A and B (see

4.26.2). Moreover, A and B are the only matrices in the pencil generated by A and B which have other than simple characteristic roots.

4.27.2 If $A = \begin{pmatrix} 0 & 0 \\ 0 & 1 \end{pmatrix}$ and $B = \begin{pmatrix} 1 & 1 \\ 0 & 1 \end{pmatrix}$, then the pencil generated by A and B has only one matrix in it which is not similar to a diagonal matrix (namely B), but A and B do not commute.

4.27.3 If

$$A = \begin{pmatrix} 0 & 1 & 0 \\ 0 & 0 & 1 \\ 0 & 0 & 0 \end{pmatrix} \quad \text{and} \quad B = \begin{pmatrix} 0 & 0 & 1 \\ 0 & 0 & 0 \\ 0 & 0 & 0 \end{pmatrix},$$

then $AB = BA$ and the pencil generated by A and B contains no matrices (other than $0_{3,3}$) similar to a diagonal matrix.

4.28   Miscellaneous results on commutativity

A number of results on commutativity with a common strand running through them are listed below.

4.28.1   Let A and B be normal matrices in $M_n(\mathbf{C})$. Let $z \in \mathbf{C}$ and let $r_i(z)$, $(i = 1, \cdots, n)$, be the characteristic roots of $A + zB$. Suppose there is an ordering of the characteristic roots of A and B, say $\alpha_1, \cdots, \alpha_n$ and $\beta_1, \cdots, \beta_n$ respectively, for which the inequality

$$\sum_{i=1}^{n} |r_i(z)|^2 \geq \sum_{i=1}^{n} |\alpha_i + z\beta_i|^2$$

holds for at least three values of z which form the vertices of a triangle with the origin in its interior. Then $AB = BA$.

4.28.2   If A and B are normal and have property L, then $AB = BA$.

4.28.3   Let $A \in M_n(\mathbf{C})$ with characteristic roots $r_1, \cdots, r_n$. If there exists a permutation $\sigma \in S_n$ and a pair of complex numbers c and h such that the characteristic roots of $cA + hA^*$ are $cr_i + hr_{\sigma(i)}$, $(i = 1, \cdots, n)$, then A is normal.

4.28.4   Let A and B be nonsingular matrices in $M_n(\mathbf{C})$. If A commutes with $AB - BA$, then 1 is a characteristic root of $A^{-1}B^{-1}AB$ of algebraic multiplicity n.

4.28.5   If $A \in M(\mathbf{C})$ commutes with $AA^* - A^*A$, then A is normal (i.e., $AA^* = A^*A$).

4.28.6   If A and B are in $M_n(\mathbf{C})$, both AB and BA are positive definite hermitian, and A commutes with $AB - BA$, then $AB = BA$. (The choice $B = A^*$ gives **4.28.5**.)

4.28.7   Let A and B be unitary matrices in $M_n(\mathbf{C})$ and set $U = ABA^{-1}B^{-1}$. If the characteristic roots of B lie on an arc of the unit circle less than a semicircle and A commutes with U, then A commutes with B.

4.28.8   Let A and B be matrices in $M_n(\mathbf{C})$ and assume that A commutes with $ABA^{-1}B^{-1}$. Then there exists a unitary $U \in M_n(\mathbf{C})$ such that $UBU^{-1} = PK$ where P is a permutation matrix and K commutes with A.

5   Congruence

5.1   Definitions

Most of the reduction theorems for matrices discussed in sections 3 and 4 give canonical forms for a matrix A under equivalence, PAQ, or similarity, PAP^{-1}. In this brief concluding section we discuss reduction theorems for a matrix with respect to the relation of *congruence*.

For the purposes of this section let $\mathbf{K}$ designate any one of the sets $\mathbf{R}$, $\mathbf{C}$, $\mathbf{I}$, $\mathbf{R}[\lambda]$, $\mathbf{C}[\lambda]$. If A and B are in $M_n(\mathbf{K})$, then A and B are *congruent over* $\mathbf{K}$, if there is a matrix $P \in M_n(\mathbf{K})$ such that $B = P^T AP$ and moreover P is nonsingular with $P^{-1} \in M_n(\mathbf{K})$. Note that matrices congruent to symmetric (skew-symmetric) matrices are themselves symmetric (skew-symmetric).

5.2   Triple diagonal form

If $A \in M_n(\mathbf{K})$ is symmetric, then A is congruent to a matrix T in *triple diagonal form*:

$$
T = \begin{pmatrix}
t_{11} & t_{12} & 0 & 0 & 0 & & & & 0 \\
t_{21} & t_{22} & t_{23} & 0 & 0 & & & & 0 \\
0 & t_{32} & t_{33} & t_{34} & 0 & & & & 0 \\
\cdot & & & & \cdot & & & & \cdot \\
\cdot & & & & & \cdot & & & \cdot \\
\cdot & & & & & & \cdot & & \cdot \\
0 & & & & 0 & t_{n-1,n-2} & t_{n-1,n-1} & t_{n-1,n} \\
0 & & & & 0 & 0 & t_{n,n-1} & t_{n,n}
\end{pmatrix}, \quad (t_{ij} = t_{ji} \text{ all } i,j).
$$

That is, T can have nonzero entries only on, directly above, or directly below the main diagonal.

5.3 Congruence and elementary operations

Let $A \in M_n(\mathbf{K})$ and suppose an elementary row operation and the corresponding elementary column operation are performed on A. The result is a matrix congruent to A. For example, if c times row i is added to row j and then c times column i is added to column j, then the resulting matrix is congruent to A. For, if E denotes the elementary matrix corresponding to an elementary row operation, then postmultiplication by E^T represents the corresponding elementary column operation.

5.4 Example

Take A to be the matrix

$$\begin{pmatrix} 1 & 2 & 4 & 0 \\ 2 & 3 & 5 & 1 \\ 4 & 5 & 2 & 0 \\ 0 & 1 & 0 & 1 \end{pmatrix} \in M_4(\mathbf{I}).$$

Perform the following sequence of congruence operations: $\mathrm{II}_{(3)-2(2)}$, $\mathrm{II}^{(3)-2(2)}$; $\mathrm{II}_{(4)+(3)}$, $\mathrm{II}^{(4)+(3)}$. This results in the matrix

$$\begin{pmatrix} 1 & 2 & 0 & 0 \\ 2 & 3 & -1 & 0 \\ 0 & -1 & -6 & -8 \\ 0 & 0 & -8 & -9 \end{pmatrix}$$

in triple diagonal form.

5.5 Relationship to quadratic forms

Let $A \in M_n(\mathbf{K})$ and let $x = (x_1, \cdots, x_n) \in M_{n,1}(\mathbf{K})$. Then $f(x) = x^T A x = \sum_{i,j=1}^{n} a_{ij} x_i x_j$ is called the *quadratic form* associated with A. Note that $f(x)$ is an element of $\mathbf{K}$ and thus $f(x) = (x^T A x)^T = x^T A^T x$ so that $f(x) = x^T B x$ where B is the symmetric matrix $(A + A^T)/2$. Thus in looking at $f(x)$ the matrix A can be taken as symmetric. If $P \in M_n(\mathbf{K})$, then $f(Px) = x^T(P^T A P)x$; therefore if $P^{-1} \in M_n(\mathbf{K})$, it follows that $f(Px) = g(x)$ in which $g(x)$ is the quadratic form associated with the matrix $P^T A P$ congruent to A.

5.6 Example

Let $f(x) = x_1^2 + 2x_2^2 + 2x_1x_2$. Then $f(x) = x^TAx$, $A = \begin{pmatrix} 1 & 1 \\ 1 & 2 \end{pmatrix} \in M_2(\mathbf{I})$. If

$P = \begin{pmatrix} 1 & -1 \\ 0 & 1 \end{pmatrix}$, then $g(x) = f(Px) = x^TBx$ in which

$$B = P^TAP = \begin{pmatrix} 1 & 0 \\ -1 & 1 \end{pmatrix}\begin{pmatrix} 1 & 1 \\ 1 & 2 \end{pmatrix}\begin{pmatrix} 1 & -1 \\ 0 & 1 \end{pmatrix} = I_2.$$

Thus $g(x) = x^TI_2x = x_1^2 + x_2^2$, and $f(x)$ takes on the same set of integer values as $x_1^2 + x_2^2$ does as x_1 and x_2 run over $\mathbf{I}$.

5.7 Congruence properties

5.7.1   If $A = A^T \in M_n(\mathbf{K})$, $\rho(A) = r$, and $\mathbf{K}$ is $\mathbf{R}$ or $\mathbf{C}$, then A is congruent over $\mathbf{K}$ to a matrix B of the form diag $(d_1, \cdots, d_r, 0, \cdots, 0) \in M_n(\mathbf{K})$ in which $d_i \neq 0$, $(i = 1, \cdots, r)$. In case $\mathbf{K} = \mathbf{C}$, all the d_i may be taken to be 1; in case $\mathbf{K} = \mathbf{R}$, all the d_i may be taken as ± 1.

Proof: By **4.10.6**, there exists a unitary matrix $U \in M_n(\mathbf{C})$ such that $U^TAU = \text{diag } (d_1, \cdots, d_r, 0, \cdots, 0)$. If $A \in M_n(\mathbf{R})$ then, by **4.10.5**, the matrix U can be chosen to be real orthogonal. In this case, let $G = \text{diag } (|d_1|^{-1/2}, \cdots, |d_r|^{-1/2}, 1, \cdots, 1)$. Then $(GU)^TA(GU)$ is a diagonal matrix whose (j,j) entry is 1 or -1 according as $d_j > 0$ or $d_j < 0$, $(j = 1, \cdots, r)$. If $A \in M_n(\mathbf{C})$, let $H = \text{diag } (d_1^{-1/2}, \cdots, d_r^{-1/2}, 1, \cdots, 1)$. Then $(HU)^TA(HU) = I_r \dot{+} 0_{n-r}$.

5.7.2   (Sylvester's law of inertia)   If

$$D = \text{diag } (d_1, \cdots, d_n) \quad \text{and} \quad E = \text{diag } (k_1, \cdots, k_n)$$

are congruent matrices over $\mathbf{R}$, then $\rho(D) = \rho(E)$ and the number of positive d_i is the same as the number of positive k_i. If A is a symmetric matrix over $\mathbf{R}$ then, according to **5.7.1**, A is congruent to a diagonal matrix D over $\mathbf{R}$. The number of positive main diagonal entries of D is called the *index* of A. The *signature* of A is the difference between the index of A and the number of negative main diagonal entries of D.

Proof: Let $E = P^TDP$. It follows from **2.17.3** that the rank of a matrix is invariant under multiplication by a nonsingular matrix. We have therefore $\rho(D) = \rho(E)$. By virtue of **5.7.1**, we assume without loss of generality that $d_j = 1$, $(j = 1, \cdots, s)$, $d_j = -1$, $(j = s + 1, \cdots, r)$; and $k_j = 1$, $(j = 1, \cdots, t)$, $k_j = -1$, $(j = t + 1, \cdots, r)$. We have to prove that $s = t$. Suppose that $s < t$. Let $x \in M_{n,1}(\mathbf{R})$ be a nonzero column vector whose last $n - t$ entries are 0 and such that the first s entries of $y = Px$ are 0. Such x can always be found since a system of homogeneous linear equations in t unknowns with $s < t$ always has a nontrivial solution (see **3.1.5**). Now,

$$x^T E x = y^T D y$$

and $x^T E x = x_{11}^2 + \cdots + x_{t1}^2 > 0$ while $y^T D y = -y_{s+1,1}^2 - \cdots - y_{t1}^2 \leq 0$, a contradiction.

5.7.3   If A and B are symmetric matrices in $M_n(\mathbf{R})$, then A and B are congruent over $\mathbf{R}$, if and only if they have the same rank and index.

5.7.4   If $A \in M_n(\mathbf{K})$ and $A^T = -A$ (A is skew-symmetric), then $\rho(A) = 2p$ and A is congruent over $\mathbf{K}$ to a matrix of the form

$$\sum_{i=1}^{p} \cdot \begin{pmatrix} 0 & q_i \\ -q_i & 0 \end{pmatrix} \dotplus 0_{n-2p,n-2p}$$

in which $q_1, q_1, q_2, q_2, \cdots, q_p, q_p$ are the invariant factors of A. Thus two skew-symmetric matrices A and B are congruent over $\mathbf{K}$, if and only if they are equivalent (see **3.17.3**) over $\mathbf{K}$.

5.8   Hermitian congruence

In case $\mathbf{K} = \mathbf{C}$ the related notion of hermitian congruence arises. If A and B are in $M_n(\mathbf{C})$, then A and B are *hermitely congruent* or *conjunctive*, if there exists a nonsingular $P \in M_n(\mathbf{C})$ such that $A = P^*BP$. Note that matrices conjunctive to hermitian (skew-hermitian) matrices are hermitian (skew-hermitian).

5.8.1   Suppose A and B are conjunctive over $\mathbf{C}$, $A = P^*BP$. Then P is a product of elementary matrices, $P = P_1 \cdots P_m$ so that

$$A = P_m^* \cdots P_1^* B P_1 \cdots P_m.$$

Thus A and B are conjunctive if and only if A can be obtained from B by a sequence of elementary row and corresponding conjunctive column operations. For example, if c times row i is added to row j and $\bar{c}$ times column i is added to column j, then the resulting matrix is hermitely congruent to the one we started with.

5.8.2   If $A = A^* \in M_n(\mathbf{C})$ and $\rho(A) = r$, then A is hermitely congruent to a matrix of the form

$$I_k \dotplus (-I_{r-k}) \dotplus 0_{n-r,n-r}$$

in which k is the *index* of A; i.e., the number of positive characteristic roots of A. The matrix A is positive semidefinite, if and only if $r = k$ and positive definite, if and only if $r = k = n$. The integer $2k - r$ is called the *signature* of A.

5.8.3   A matrix $A \in M_n(\mathbf{C})$ is positive definite hermitian, if and only if

$A = P^*P$ for some nonsingular $P \in M_n(\mathbf{C})$. In other words, A is positive definite hermitian if and only if A is hermitely congruent to I_n. For, if A is positive definite hermitian, then (see **4.10.3**) there exists a unitary matrix U such that $UAU^* = D$ where $D = \text{diag}\,(r_1^2, \cdots, r_n^2)$. Let $F = \text{diag}\,(r_1, \cdots, r_n)$ and $P = U^*FU$. Then

$$P^*P = (U^*FU)^2 = U^*F^2U = U^*DU = A.$$

Since A is nonsingular, the matrix P must be nonsingular.

Conversely, if $A = P^*P$, then clearly A is hermitian and thus there exists a unitary matrix U such that $UP^*PU^* = G = \text{diag}\,(g_1, \cdots, g_n)$. But $G = (UP^*)(UP^*)^*$ and therefore $g_j = ||(UP^*)^{(j)}|| > 0$, $(j = 1, \cdots, n)$.

5.8.4 A matrix $A \in M_n(\mathbf{C})$, $\rho(A) = r$, is positive semidefinite hermitian, if and only if A is hermitely congruent to $I_r \dotplus 0_{n-r,n-r}$.

5.9 Example

Let $A = \begin{pmatrix} 1 & 1 \\ 1 & 2 \end{pmatrix}$. Then if $Q = \begin{pmatrix} 1 & 0 \\ -1 & 1 \end{pmatrix}$, $I_2 = Q^TAQ$, so that $A = (Q^{-1})^TQ^{-1} = P^*P$ where $P = \begin{pmatrix} 1 & 0 \\ 1 & 1 \end{pmatrix}$.

5.10 Triangular product representation

The result of **5.8.3** can be improved by specifying that P be triangular.

5.10.1 If $A \in M_n(\mathbf{C})$, then A is positive definite hermitian, if and only if $A = P^*P$ in which P is a nonsingular upper triangular matrix (i.e., $P_{ij} = 0$ for $i > j$).

5.10.2 In order to compute the upper triangular matrix P in **5.10.1**, the Gram-Schmidt process (see **4.5.2**) is used.

Step 1. By a sequence of elementary row operations $E_1, \cdots, E_m$ and corresponding conjunctive column operations (see **5.8.1**) reduce A to I_n; i.e., $E_m^* \cdots E_1^*AE_1 \cdots E_m = I_n$. Then $A = Q^*Q$ where $Q = (E_1 \cdots E_m)^{-1}$.

Step 2. The columns $Q^{(1)}, \cdots, Q^{(n)}$ are linearly independent (Q is nonsingular) so that, by **4.5.2**, there exists an orthonormal set of n-tuples $u_1, \cdots, u_n$ such that

$$
\begin{aligned}
Q^{(1)} &= c_{11}u_1 \\
Q^{(2)} &= c_{12}u_1 + c_{22}u_2 \\
&\ \ \vdots \\
Q^{(n)} &= c_{1n}u_1 + \cdots + c_{nn}u_n.
\end{aligned}
$$

Step 3.   Let U be the unitary matrix (see **4.7.24**) whose columns are the vectors $u_1, \cdots, u_n$ in order. Then let P be the upper triangular matrix

$$\begin{pmatrix} c_{11} & c_{12} & c_{13} & \cdots & c_{1n} \\ 0 & c_{22} & c_{23} & \cdots & c_{2n} \\ 0 & 0 & c_{33} & \cdots & c_{3n} \\ \cdot & \cdot & 0 & \cdot & \cdot \\ \cdot & \cdot & \cdot & \cdot & \cdot \\ \cdot & \cdot & \cdot & & \cdot \cdot \\ 0 & 0 & 0 & \cdots & c_{nn} \end{pmatrix}.$$

Thus $Q = UP$ so that $A = Q^*Q = P^*U^*UP = P^*P$. (Recall that $U^*U = I_n$ because U is unitary.)

5.11 Example

Let

$$A = \begin{pmatrix} 9 & 5 & 2 \\ 5 & 3 & 1 \\ 2 & 1 & 1 \end{pmatrix}.$$

Then, using the notation of **5.10.2**, we have $A = Q^*Q$ where

$$Q = \begin{pmatrix} 1 & 1 & 0 \\ -2 & -1 & 0 \\ 2 & 1 & 1 \end{pmatrix}.$$

Now apply the Gram-Schmidt process (see **4.5.2**) to obtain

$$Q^{(1)} = 3u_1,$$

$$Q^{(2)} = \frac{5}{3} u_1 + \frac{\sqrt{2}}{3} u_2,$$

$$Q^{(3)} = \frac{2}{3} u_1 - \frac{\sqrt{2}}{6} u_2 + \frac{1}{\sqrt{2}} u_3,$$

where

$$u_1 = \left(\frac{1}{3}, -\frac{2}{3}, \frac{2}{3}\right), \quad u_2 = \left(\frac{4}{3\sqrt{2}}, \frac{1}{3\sqrt{2}}, -\frac{1}{3\sqrt{2}}\right), \quad u_3 = \left(0, \frac{1}{\sqrt{2}}, \frac{1}{\sqrt{2}}\right)$$

form an orthonormal set. Hence

$$P = \begin{pmatrix} 3 & \dfrac{5}{3} & \dfrac{2}{3} \\ 0 & \dfrac{\sqrt{2}}{3} & -\dfrac{\sqrt{2}}{6} \\ 0 & 0 & \dfrac{1}{\sqrt{2}} \end{pmatrix}$$

is the upper triangular matrix satisfying $P^*P = A$.

5.12 Conjunctive reduction of skew-hermitian matrices

If $A \in M_n(\mathbf{C})$ and $A^* = -A$ then $H = -iA$ is hermitian and, by **5.8.2**, there exists a nonsingular $P \in M_n(\mathbf{C})$ such that

$$-iP^*AP = P^*HP = I_k \dotplus (-I_{r-k}) \dotplus 0_{n-r,n-r}$$

where $r = \rho(A)$ and k is the index of $-iA$. Thus A is hermitely congruent to

$$iI_k \dotplus (-iI_{r-k}) \dotplus 0_{n-r,n-r}.$$

5.13 Conjunctive reduction of two hermitian matrices

If A and B are n-square hermitian matrices and A is positive definite hermitian, then there exists a nonsingular $P \in M_n(\mathbf{C})$ such that

$$P^*AP = I_n,$$
$$P^*BP = \text{diag }(k_1, \cdots, k_n)$$

in which k_i, $(i = 1, \cdots, n)$, are the characteristic roots of $A^{-1}B$. If A and B are in $M_n(\mathbf{R})$, then P may be chosen in $M_n(\mathbf{R})$. Let $A^{1/2}$ be the positive definite hermitian square root of A (see **4.14.1**). Then

$$\lambda(A^{-1}B) = \lambda(A^{1/2}(A^{-1}B)A^{-1/2}) = \lambda(A^{-1/2}BA^{-1/2}).$$

The matrix $A^{-1/2}BA^{-1/2}$ is hermitian and hence (see **4.7.18**) has real characteristic roots. Thus the numbers $k_1, \cdots, k_n$ are real.

References

Some of the material in this chapter can be found in many of the books listed in Part A of the Bibliography that follows. No explicit references are given in such cases. In other cases, however, a reference is given by a number preceded by A or by B according to whether the work referred to is a book or a research paper. Separate reference lists and bibliographies are provided for each chapter.

§1.9.   [A27], p. 81.
§1.10.   *Ibid.*, p. 89; [B6]; [B15].
§2.4.10.   [B16].
§2.4.14.   [A11], vol. I, p. 9; for the history of this theorem and the respective claims by Cauchy and Binet see [A29], vol. I, pp. 118–131.

§2.5.   [A18], p. 309.

§2.7.   [A27], p. 86; [A48], p. 64.

§2.7.2.   The Sylvester-Franke theorem, as stated here, is due to Sylvester. Franke generalized this result to a theorem on a minor of the rth compound matrix ([A46], p. 108). Actually the Sylvester-Franke theorem given here is virtually due to Cauchy ([A29], vol. I, p. 118).

§§2.9–2.11.   [B17].

§2.12.   [A27], p. 81.

§2.12.7.   [B11].

§2.15.11.   [A27], p. 84.

§2.15.12.   *Ibid.*, p. 87.

§2.15.14.   *Ibid.*, p. 86.

§2.15.16.   [A5], p. 246.

§2.15.17.   *Ibid.*, p. 252.

§2.16.2.   [A17], p. 94.

§2.16.3.   [B5].

§2.17.1.   [A27], p. 11.

§3.28.2.   *Ibid.*, p. 18.

§3.32.5.   [B4].

§4.9.   [A17], p. 94.

§4.10.2.   [A27], p. 76.

§4.10.6.   [B18].

§4.10.7.   [B8].

§4.12.2.   [A2], p. 134.

§4.12.3.   *Ibid.*, p. 134.

§§4.12.4, 4.12.5.   [A47], pp. 91–93.

§4.12.7.   [A48], ch. V; [B17].

§4.17.   [A48], p. 116; [A11], pp. 98–100.

§4.19.1.   [A11], vol. I, p. 278.

§4.19.2.   *Ibid.*, p. 279.

§4.19.3.   [A11], vol. II, p. 4.

§4.19.4.   [A16], p. 169; [A11], vol. I, p. 276.

§4.19.5.   [A11], vol. II, p. 4.

§4.19.6.   *Ibid.*, p. 6.

§§4.19.7, 4.19.8.   [B1].

§4.19.9.   [B10].

§§4.20.2, 4.20.3.   [B2].

§4.21.3.   [A11], vol. I, p. 292.

§4.22.2.   [B3].

§4.22.3.   [A5], p. 241; [A11], vol. I, p. 222.

§4.22.4.   [A48], p. 106.

§§4.24.1–4.24.3.   [B3].

§§4.25, 4.26.   [B19].

§§4.26.1–4.26.3.   *Ibid.*

§§4.27.1–4.27.3.   *Ibid.*

§4.28.1.   [B23].

§4.28.2.   [B22].

§4.28.3. [B7].
§4.28.4. [B20].
§4.28.5. [B9].
§4.28.6. [B20].
§4.28.7. [B21].
§4.28.8. [B21].
§5.2. [B14], p. 246

Bibliography

Part A. Books

1. Aitken, A. C., *Determinants and matrices*, 4th ed., Oliver and Boyd, Edinburgh (1946).
2. Amir-Moéz, A. R., and Fass, A. L., *Elements of linear spaces*, Pergamon, New York (1962).
3. Bellman, R., *Introduction to matrix analysis*, McGraw-Hill, New York (1960).
4. Bodewig, E., *Matrix calculus*, 2nd ed., North-Holland, Amsterdam (1959).
5. Browne, E. T., *Introduction to the theory of determinants and matrices*, University of North Carolina, Chapel Hill (1958).
6. Faddeeva, V. N., *Computational methods of linear algebra*, Dover, New York (1958).
7. Ferrar, W. L., *Finite matrices*, Clarendon, Oxford (1951).
8. Finkbeiner, D. T., *Introduction to matrices and linear transformations*, Freeman, San Francisco (1960).
9. Frazer, R. A., Duncan, W. J., and Collar, A. R., *Elementary matrices*, Cambridge University, London (1938).
10. Fuller, L. E., *Basic matrix theory*, Prentice-Hall, Englewood Cliffs, N.J. (1962).
11. Gantmacher, F. R., *The theory of matrices*, vols. I and II (trans., K. A. Hirsch), Chelsea, New York (1959).
12. Gelfand, I. M., *Lectures on linear algebra*, Interscience, New York (1961).
13. Graeub, W., *Lineare Algebra*, Springer, Berlin (1958).
14. Gröbner, W., *Matrizenrechnung*, Oldenburg, Munich (1956).
15. Hadley, G., *Linear algebra*, Addison-Wesley, Reading, Mass. (1961).
16. Halmos, P. R., *Finite dimensional vector spaces*, 2nd ed., Van Nostrand, Princeton (1958).
17. Hamburger, H. L., and Grimshaw, M. E., *Linear transformations in n-dimensional vector space*, Cambridge University, London (1951).
18. Hodge, W. V. D., and Pedoe, D., *Methods of algebraic geometry*, vol. I, Cambridge University, London (1947).

19. Hoffman, K., and Kunze, R., *Linear algebra*, Prentice-Hall, Englewood Cliffs, N.J. (1961).

20. Hohn, F. E., *Elementary matrix algebra*, Macmillan, New York (1958).

21. Jacobson, N., *Lectures in abstract algebra*, vol. II, Van Nostrand, New York (1953).

22. Jaeger, A., *Introduction to analytic geometry and linear algebra*, Holt, Rinehart and Winston, New York (1960).

23. Jung, H. W. E., *Matrizen und Determinanten*, Fachbuchverlag, Leipzig (1952).

24. Kowalewski, G., *Einführung in die Determinantentheorie*, Chelsea, New York (1948).

25. Kuiper, N. H., *Linear algebra and geometry*, North-Holland, Amsterdam (1962).

26. MacDuffee, C. C., *Vectors and matrices*, Mathematical Association of America, Menasha, Wisc. (1943).

27. MacDuffee, C. C., *The theory of matrices*, Chelsea, New York (1946).

28. Mirsky, L., *An introduction to linear algebra*, Oxford University, Oxford (1955).

29. Muir, T., *Theory of determinants*, vols. I–IV, Dover, New York (1960).

30. Muir, T., *Contributions to the history of determinants 1900–1920*, Blackie, London (1930).

31. Muir, T., and Metzler, W. H., *A treatise on the theory of determinants*, Dover, New York (1960).

32. Murdoch, D. C., *Linear algebra for undergraduates*, Wiley, New York (1957).

33. Neiss, F., *Determinanten und Matrizen*, 4th ed., Springer, Berlin (1955).

34. Nering, E. D., *Linear algebra and matrix theory*, Wiley, New York (1963).

35. Paige, L. J., and Swift, J. D., *Elements of linear algebra*, Ginn, Boston (1961).

36. Parker, W. V., and Eaves, J. C., *Matrices*, Ronald, New York (1960).

37. Perlis, S., *Theory of matrices*, Addison-Wesley, Reading, Mass. (1952).

38. Pham, D., *Techniques du calcul matriciel*, Dunod, Paris (1962).

39. Schreier, O., and Sperner, E., *Modern algebra and matrix theory*, Chelsea, New York (1951).

40. Schwartz, J. T., *Introduction to matrices and vectors*, McGraw-Hill, New York (1961).

41. Schwerdtfeger, H., *Introduction to linear algebra and the theory of matrices*, Noordhoff, Groningen (1950).

42. Shilov, G. E., *An introduction to the theory of linear spaces*, Prentice-Hall, Englewood Cliffs, N.J. (1961).

43. Smirnov, V. I. (rev. by R. A. Silverman), *Linear algebra and group theory*, McGraw-Hill, New York (1961).

44. Stoll, R. R., *Linear algebra and matrix theory*, McGraw-Hill, New York (1952).

45. Thrall, R. M., and Tornheim, L., *Vector spaces and matrices*, Wiley, New York (1957).

46. Turnbull, H. W., *The theory of determinants, matrices and invariants*, 3rd ed., Dover, New York (1960).

47. Turnbull, H. W., and Aitken, A. C., *An introduction to the theory of canonical matrices*, 3rd ed., Dover, New York (1961).

48. Wedderburn, J. H. M., *Lectures on matrices*, American Mathematical Society, Providence, R.I. (1934).

49. Zurmühl, R., *Matrizen, eine Darstellung für Ingenieure*, 2nd ed., Springer, Berlin (1957).

Part B. Papers

1. de Bruijn, N. G., and Szekeres, G., *On some exponential and polar representatives of matrices*, Nieuw Arch. Wisk., *3* (1955), 20–32.
2. Diliberto, S. P., *On stability of linear mechanical systems*, Office of Naval Research Technical Report, NONR 222(88), University of California, Berkeley (May, 1962).
3. Drazin, M. A., Dungey, J. W., and Gruenberg, K. W., *Some theorems on commutative matrices*, J. London Math. Soc., *26* (1951), 221–228.
4. Flanders, H., and Epstein, M., *On the reduction of a matrix to diagonal form*, Amer. Math. Monthly, *62* (1955), 168.
5. Friedman, B., *Eigenvalues of composite matrices*, Proc. Cambridge Philos. Soc., *57* (1961), 37–49.
6. Givens, W., *Elementary divisors and some properties of the Lyapunov mapping* $X \rightarrow AX + XA^*$, Argonne National Laboratory Report, ANL-6456 (1962), 11.
7. Hoffman, A. J., and Taussky, O., *A characterization of normal matrices*, J. Res. Nat. Bur. Standards, *52* (1954), 17–19.
8. Hua, L. K., *On the theory of automorphic functions of a matrix variable I— geometrical basis*, Amer. J. Math., *66* (1944), 470–488.
9. Kato, T., and Taussky, O., *Commutators of A and A**, J. Washington Acad. Sci., *46* (1956), 38–40.
10. Marcus, M., *Matrices in linear mechanical systems*, Canad. Math. Bull., *5* (1962), 253–257.
11. Marcus, M., and Newman, M., *Inequalities for the permanent function*, Ann. of Math., *75* (1962), 47–62.
12. Motzkin, T. S., and Taussky, O., *Pairs of matrices with property L*, Trans. Amer. Math. Soc., I, *73* (1952) 108–114; II, *80* (1955), 387–401.
13. Motzkin, T. S., and Taussky, O., *Pairs of matrices with property L*, Proc. Nat. Acad. Sci., *39* (1953), 961–963.
14. Newman, M., *Matrix computations*. (An article in *A survey of numerical analysis*, ed. by J. Todd, McGraw-Hill, New York, 1962.)
15. Roth, W. E., *On direct product matrices*, Bull. Amer. Math. Soc., *40* (1934), 461–468.
16. Ryser, H. J., *Geometries and incidence matrices*, Slaught Paper No. 4, Mathematical Association of America (1955).
17. Ryser, H. J., *Compound and induced matrices in combinatorial analysis*, Proc. of Symposia in Appl. Math., vol. 10, American Mathematical Society (1960), 139–167.
18. Schur, I., *Ein Satz ueber quadratische Formen mit komplexen Koeffizienten*, Amer. J. Math., *67* (1945), 472.
19. Taussky, O., *Commutativity in finite matrices*, Amer. Math. Monthly, *64* (1957), 229–235.

20. Taussky, O., *A note on the group commutator of* A *and* A*, J. Washington Acad. Sci., *48* (1958), 305.

21. Taussky, O., *Commutators of unitary matrices which commute with one factor*, J. Math. Mech., *10* (1961), 175–178.

22. Wiegmann, N. A., *A note on pairs of normal matrices with property* L, Proc. Amer. Math. Soc., *4* (1953), 35–36.

23. Wielandt, H., *Pairs of normal matrices with property* L, J. Res. Nat. Bur. Standards, *51* (1953), 89–90.

Convexity and Matrices

II

1 Convex Sets

1.1 Definitions

In this chapter we shall be mainly interested in subsets of $\mathbf{R}^n$ and $M_{m,n}(\mathbf{R})$. Some of the statements that follow are true for both $\mathbf{R}^n$ and $M_{m,n}(\mathbf{R})$, in which case we shall designate either of the sets by $\mathbf{U}$.

The inner product used in $\mathbf{R}^n$ is the one defined in **I.4.4**: if $u = (u_1, \cdots, u_n) \in \mathbf{R}^n$ and $v = (v_1, \cdots, v_n) \in \mathbf{R}^n$, then $(u, v) = \sum_{i=1}^{n} u_i v_i$. The inner product used for $M_{m,n}(\mathbf{R})$ is $(A, B) = \operatorname{tr}(AB^T)$.

If x and y are elements in $\mathbf{U}$ and θ is a real number, then the set

$$\{z \in \mathbf{U} \mid z = \theta x + (1 - \theta)y, \qquad 0 \leq \theta \leq 1\} \tag{1}$$

is called the *line segment* joining x and y. If with any two elements of a subset S of $\mathbf{U}$ the line segment joining them is a subset of S, we say that the set S is *convex*. For example, the following sets are clearly convex:

(i)   a set consisting of a single element,
(ii) a line segment,
(iii) a subspace of $\mathbf{U}$.

On the other hand the following sets are not convex:

(iv)   a set of a finite number, p, $(p > 1)$, of elements,
(v)   the complement in $\mathbf{U}$ of a line segment,
(vi)   a set-theoretical union of two subspaces of $\mathbf{U}$ neither of which is a
subspace of the other. For, if S_1 and S_2 are two such subspaces,
then there exist elements x and y such that $x \in S_1$ and $x \notin S_2$,
$y \in S_2$ and $y \notin S_1$. If $S_1 \cup S_2$ were convex, then $z = (x + y)/2$
would belong to $S_1 \cup S_2$ and therefore either to S_1 or to S_2. But if
$z \in S_1$, then $2z - x = y \in S_1$, which contradicts our assumptions.
Similarly, $z \in S_2$ leads to a contradiction.

Let x and y be two distinct, fixed elements of $\mathbf{U}$. The set

$$\{z \in \mathbf{U} | z = \theta x + (1 - \theta)y, \qquad \theta \in \mathbf{R}\} \tag{2}$$

is called a *line* through x, y. Any line is convex.

Let z be a fixed nonzero element of $\mathbf{U}$ and α a fixed real number. The set

$$\{x \in \mathbf{U} | (x, z) = \alpha\} \tag{3}$$

is called a *hyperplane* in $\mathbf{U}$. Alternative definition: a hyperplane is the set

$$\{x \in \mathbf{U} | x = z + y, \qquad y \in \langle z \rangle^\perp\} \tag{4}$$

where $\langle z \rangle^\perp$ denotes the orthogonal complement of $\langle z \rangle$ in $\mathbf{U}$; i.e., the space
of all vectors u in $\mathbf{U}$ such that $(u, z) = 0$. A hyperplane is convex.

If α is a fixed real number, the sets

$$\{x \in \mathbf{U} | (x, z) < \alpha\} \qquad \text{and} \qquad \{x \in \mathbf{U} | (x, z) > \alpha\} \tag{5}$$

are called *open half-spaces*, while the sets

$$\{x \in \mathbf{U} | (x, z) \leq \alpha\} \qquad \text{and} \qquad \{x \in \mathbf{U} | (x, z) \geq \alpha\} \tag{6}$$

are *closed half-spaces*. Half-spaces, open or closed, are convex. Thus a
hyperplane divides $\mathbf{U}$ into three disjoint complementary convex sets: the
hyperplane itself and two open half-spaces.

1.2   Examples

1.2.1   The set D_n of real n-square matrices whose row sums and column
sums are all equal to 1 clearly form a convex set in $M_n(\mathbf{R})$. For, if $A = (a_{ij})$,
$B = (b_{ij})$ and

$$\sum_{j=1}^{n} a_{ij} = \sum_{i=1}^{n} a_{ij} = \sum_{j=1}^{n} b_{ij} = \sum_{i=1}^{n} b_{ij} = 1,$$

then, for any θ,

$$\sum_{j=1}^{n} (\theta a_{ij} + (1 - \theta)b_{ij}) = \theta \sum_{j=1}^{n} a_{ij} + (1 - \theta) \sum_{j=1}^{n} b_{ij}$$
$$= \theta + (1 - \theta) = 1.$$

Similarly, the column sums of $\theta A + (1 - \theta)B$ are all equal to 1. If

$J \in M_n(\mathbf{R})$ denotes the matrix all of whose entries are equal to 1, then $A \in D_n$, if and only if $JA = AJ = J$.

1.2.2   The set of all matrices in $M_n(\mathbf{R})$ each of whose entries is nonnegative is convex.

1.3   Intersection property

The intersection of the members of any family of convex subsets of $\mathbf{U}$ is convex.

1.4   Examples

1.4.1   The set of all vectors of $\mathbf{R}^n$ with nonnegative coordinates (the *nonnegative orthant* of $\mathbf{R}^n$) is convex. For, this set is the intersection of n closed half-spaces

$$\{z \in \mathbf{R}^n | (z, e_k) \geq 0, \qquad k = 1, \cdots, n\},$$

where e_k denotes the n-tuple whose kth coordinate is 1 and whose other $n - 1$ coordinates are 0. (See **I.2.19**.)

1.4.2   A matrix $A \in M_n(\mathbf{R})$ is called *doubly stochastic*, if all its entries are nonnegative and all its row sums and all its column sums are equal to 1. The set of all n-square doubly stochastic matrices, Ω_n, is convex, since it is the intersection of D_n (see **1.2.1**) and the set in **1.2.2**.

1.4.3   Let $W \in M_n(\mathbf{C})$ be a unitary matrix. Then the matrix $S \in M_n(\mathbf{R})$ whose (i,j) entry is $|W_{ij}|^2$ is doubly stochastic. For, from **I.4.7.24**, the rows and columns of W are unit vectors. Such a doubly stochastic matrix is called *orthostochastic*.

1.4.4   The matrix

$$S = \tfrac{1}{2} \begin{pmatrix} 1 & 1 & 0 \\ 1 & 0 & 1 \\ 0 & 1 & 1 \end{pmatrix}$$

is doubly stochastic but not orthostochastic. This is easily seen as follows. If W were a unitary matrix for which $|W_{ij}|^2 = s_{ij}$, then W would have the form

$$W = \begin{pmatrix} x_1 & y_1 & 0 \\ x_2 & 0 & y_2 \\ 0 & x_3 & y_3 \end{pmatrix}.$$

But $x_1 \bar{x}_2 = 0$, by **I.4.7.24**, and this is incompatible with $|x_1|^2 = |x_2|^2 = \tfrac{1}{2}$.

1.5 Convex polyhedrons

A linear combination $\sum_{j=1}^{p} \theta_j x_j$ of elements of $\mathbf{U}$ is called a *convex combination*
or a *convex sum* of $x_1, \cdots, x_p$, if all the θ_j are nonnegative and $\sum_{j=1}^{p} \theta_j = 1$.
The set of all such convex combinations of $x_1, \cdots, x_p$ is called the *convex*
polyhedron spanned by $x_1, \cdots, x_p$ and will be denoted by $H(x_1, \cdots, x_p)$.
More generally, if $X \subset \mathbf{U}$, then $H(X)$ will denote the totality of convex
combinations of finite sets of elements of X. The set $H(X)$ is called the
convex hull of X. A convex polyhedron is a convex set.

Some elementary properties of the convex hull are:

(i)   if $X \subset Y$, then $H(X) \subset H(Y)$;
(ii) for any X, $H(H(X)) = H(X)$;
(iii) if $x \in H(x_1, \cdots, x_p)$, then
      $H(x, x_1, \cdots, x_p) = H(x_1, \cdots, x_p)$.

1.5.1   There exists a unique finite set of elements $y_1, \cdots, y_r$, $y_j \in$
$H(x_1, \cdots, x_p)$, $(j = 1, \cdots, r)$, such that $y_j \notin H(y_1, \cdots, y_{j-1}, y_{j+1}, \cdots, y_r)$
and $H(y_1, \cdots, y_r) = H(x_1, \cdots, x_p)$.

We prove this statement in two parts. The existence of the set $y_1, \cdots, y_r$
is established by induction on the integer p. There is nothing to prove
for $p = 1$. If no x_j is a convex combination of x_i, $(i \neq j)$, then $x_1, \cdots, x_p$
will do for the choice of the y's. Otherwise we can assume without loss of
generality that $x_1 \in H(x_2, \cdots, x_p)$. By induction there exist $y_1, \cdots, y_r$ such
that no y_j is a convex combination of y_i, $(i \neq j)$; and moreover,
$H(y_1, \cdots, y_r) = H(x_2, \cdots, x_p)$. But $x_1 \in H(x_2, \cdots, x_p) = H(y_1, \cdots, y_r)$ so
that $H(y_1, \cdots, y_r) = H(x_1, \cdots, x_p)$.

To prove the uniqueness suppose that $u_1, \cdots, u_s$ are in $H(x_1, \cdots, x_p)$, no
u_j is a convex combination of the u_i, $(i \neq j)$, and $H(u_1, \cdots, u_s) =$
$H(x_1, \cdots, x_p) = H(y_1, \cdots, y_r)$. Then $y_1 = \sum_{j=1}^{s} \alpha_j u_j$, $\sum_{j=1}^{s} \alpha_j = 1$, $\alpha_j \geq 0$ and
$u_j = \sum_{t=1}^{r} \beta_{jt} y_t$, $\sum_{t=1}^{r} \beta_{jt} = 1$, $\beta_{jt} \geq 0$, $(j = 1, \cdots, s; t = 1, \cdots, r)$. We com-
pute that $y_1 = \sum_{t=1}^{r} \left(\sum_{j=1}^{s} \alpha_j \beta_{jt} \right) y_t$ and hence $\sum_{j=1}^{s} \alpha_j \beta_{j1} = 1$, $\sum_{j=1}^{s} \alpha_j \beta_{jt} = 0$,
$(t = 2, \cdots, r)$. It follows that $1 = \sum_{j=1}^{s} \alpha_j \beta_{j1} \leq \max_{j=1, \cdots, s} \beta_{j1} \leq 1$. Thus for
some k, $1 \leq k \leq s$, $\beta_{k1} = 1$ and therefore $\beta_{kt} = 0$, $(t = 2, \cdots, r)$. Thus
$y_1 = u_k$. Similarly, we prove that $\{y_2, \cdots, y_r\}$ is a subset of $\{u_1, \cdots, u_s\}$
and reversing the roles of the u's and y's completes the argument.

1.5.2   The unique elements $y_1, \cdots, y_r$ in **1.5.1** are called the *vertices* of the polyhedron $H(x_1, \cdots, x_p)$.

1.6   Example

The set of real n-tuples

$$\{(\alpha_1, \cdots, \alpha_n) | \alpha_j \geq 0, \qquad \sum_{j=1}^{n} \alpha_j = 1\}$$

forms a convex polyhedron whose vertices are $e_1, \cdots, e_n$ (see **1.4.1**). For,

$$(\alpha_1, \cdots, \alpha_n) = \sum_{j=1}^{n} \alpha_j e_j$$

and clearly no e_j is a convex combination of the remaining e_i, $(i \neq j)$. The convexity of this set follows also from the fact that it is an intersection of a hyperplane with the nonnegative orthant.

1.7   Birkhoff theorem

The set of all n-square doubly stochastic matrices, Ω_n, forms a convex polyhedron with the permutation matrices as vertices.

In order to prove **1.7** we first prove the following lemma.

1.7.1   (Frobenius-König theorem)   Every diagonal of an n-square matrix A contains a zero element if and only if A has an $s \times t$ zero submatrix with $s + t = n + 1$.

To prove the sufficiency assume that $A[\omega|\tau] = 0_{s,t}$ where $\omega \in Q_{s,n}$, $\tau \in Q_{t,n}$. Suppose that there is a diagonal f of A without a zero element. Then in columns τ the entries of f must lie in $A(\omega|\tau) \in M_{n-s,t}(\mathbf{R})$, and hence $t \leq n - s$. But $t = n - s + 1$, a contradiction.

To prove the necessity we use induction on n. If $A = 0_{n,n}$, there is nothing to prove. Assume then that $a_{ij} \neq 0$ for some i, j. Then every diagonal of $A(i|j)$ must contain a zero element and, by the induction hypothesis, $A(i|j)$ contains a $u \times v$ zero submatrix with $u + v = (n - 1) + 1$; i.e., a $u \times (n - u)$ zero submatrix. We permute the rows and the columns of A so that if the resulting matrix is denoted by B, then $B[1, \cdots, u | u + 1, \cdots, n] = 0_{u,n-u}$; i.e., so that B is of the form $B = \begin{pmatrix} X & 0 \\ Y & Z \end{pmatrix}$ where $X \in M_u(\mathbf{R})$ and $Z \in M_{n-u}(\mathbf{R})$. If all elements of any diagonal of X are nonzero, then all diagonals of Z must contain a zero element. It follows that either all the diagonals of X or all those of Z contain a zero element. Suppose that the former is the case. Then, by the induction hypothesis, X must contain a $p \times q$ zero submatrix with $p + q = u + 1$. But then the first u rows of B contain a $p \times (q + v)$ zero submatrix and $p + (q + v) = (p + q) + n - u = u + 1 + n - u = n + 1$. If every diagonal of Z contains a zero element, the proof is almost identical.

1.7.2 We are now in position to prove the Birkhoff theorem. Let $S \in \Omega_n$. We use induction on the number of positive entries in S. The theorem holds if S has exactly n positive entries; for then S is a permutation matrix. Suppose that S is not a permutation matrix. Then S has a diagonal with positive entries. For, suppose that S contains a zero submatrix $S[\omega|\tau]$, $\omega \in Q_{s,n}$, $\tau \in Q_{t,n}$. Since the only nonzero elements in the rows ω of S are in $S[\omega|\tau)$, each row sum of $S[\omega|\tau)$ is 1 and the sum of all elements in $S[\omega|\tau)$ is s. Similarly, the sum of all elements in $S(\omega|\tau]$ is t. Now, $S[\omega|\tau)$ and $S(\omega|\tau]$ are disjoint submatrices of S and the sum of all elements in S is n. Therefore $s + t \leq n$ and, by **1.7.1**, S contains a diagonal all of whose entries are positive. Let P be the permutation matrix with 1's in the positions corresponding to this diagonal and let θ be the least element in that diagonal. Clearly $\theta < 1$ and $T = (S - \theta P)(1 - \theta)^{-1}$ belongs to Ω_n and has at least one less positive entry than S. Hence, by the induction hypothesis, T is a convex combination of permutation matrices and therefore $S = \theta P + (1 - \theta)T$ is a convex combination of permutation matrices; i.e., Ω_n is a convex polyhedron spanned by permutation matrices. Clearly no permutation matrix is a convex combination of other permutation matrices and therefore the permutation matrices are the vertices of the polyhedron Ω_n (see **1.5.1**).

1.8 Simplex

The polyhedron $H(x_0, x_1, \cdots, x_p)$, in which $x_1 - x_0,\ x_2 - x_0,\ \cdots, x_p - x_0$ are linearly independent, is called a *p-simplex*; the vectors $x_0, \cdots, x_p$ form the *basis* of the p-simplex. It is obvious that the choice of the vector x_0 is immaterial; i.e., $x_j - x_r$, $(j \neq r; j = 0, \cdots, p)$, are linearly independent, if and only if $x_j - x_0$, $(j = 1, \cdots, p)$, are linearly independent: for, $x_j - x_r = (x_j - x_0) - (x_r - x_0)$. Any element of a simplex can be uniquely represented as a convex combination of its basis elements. For, suppose that

$$\sum_{i=1}^{p} \alpha_i x_i = \sum_{i=1}^{p} \beta_i x_i, \quad \sum_{i=1}^{p} \alpha_i = \sum_{i=1}^{p} \beta_i = 1.$$

Then

$$0 = \sum_{i=1}^{p} (\alpha_i - \beta_i)x_i = \sum_{i=1}^{p} (\alpha_i - \beta_i)(x_i - x_0)$$

since

$$\sum_{i=1}^{p} (\alpha_i - \beta_i) = 0.$$

The linear independence of the vectors $x_i - x_0$, $(i = 1, \cdots, p)$, implies that $\alpha_i = \beta_i$, $(i = 1, \cdots, p)$. It is also clear that $x_0, \cdots, x_p$ are the vertices of $H(x_0, \cdots, x_p)$. In order to see this we must show that no x_j is a convex combination of x_i, $(i \neq j)$. Suppose

$$x_j = \sum_{i=0, i \neq j}^{p} \theta_i x_i, \qquad \left(\sum_{i=0, i \neq j}^{p} \theta_i = 1 \right).$$

Then

$$\sum_{i=0, i \neq j}^{p} \theta_i (x_i - x_j) = 0$$

contradicting the previous remark that $x_i - x_j$, $(i \neq j)$, are linearly independent.

1.9   Examples

1.9.1   The set described in **1.6** is an $(n-1)$-simplex with basis $e_1, \cdots, e_n$.

1.9.2   The set Ω_n, $n > 2$, is not a simplex in $M_n(\mathbf{R})$. For it is clear that if $P_1, \cdots, P_N$, $N = n!$, are all the n-square permutation matrices, then $P_i \notin H(P_1, \cdots, P_{i-1}, P_{i+1}, \cdots, P_N)$, $(i = 1, \cdots, N)$. Hence from the uniqueness of the vertices of $H(P_1, \cdots, P_N)$ (see **1.5.1**), it suffices to prove that $P_j - P_1$, $(j = 2, \cdots, N)$, are linearly dependent. But $\dim M_n(\mathbf{R}) = n^2$ (see **I.2.20.9**) and it follows that Ω_n can have no more than n^2 linearly independent matrices in it. However, $n! - 1 > n^2$ for $n \geq 4$ so $P_j - P_1$, $(j = 2, \cdots, N)$, are linearly dependent. In case $n = 3$, take

$$P_1 = I_3, \quad P_2 = \begin{pmatrix} 0 & 1 & 0 \\ 0 & 0 & 1 \\ 1 & 0 & 0 \end{pmatrix}, \quad P_3 = P_2^2, \quad P_4 = \begin{pmatrix} 0 & 0 & 1 \\ 0 & 1 & 0 \\ 1 & 0 & 0 \end{pmatrix},$$

$$P_5 = P_2 P_4, \quad P_6 = P_3 P_4.$$

Then $P_1 + P_2 + P_3 = P_4 + P_5 + P_6$ and therefore $(P_4 - P_1) + (P_5 - P_1) + (P_6 - P_1) - (P_2 - P_1) - (P_3 - P_1) = 0_{3,3}$. It follows that Ω_3 is not a simplex.

1.10   Dimension

The dimension of a p-simplex is p. The dimension of a convex set is the largest of the dimensions of the simplices it contains.

1.11   Example

We compute the dimension of Ω_n. The polyhedron Ω_n lies in the intersection of the hyperplanes

$$\sum_{j=1}^{n} x_{ij} = 1, \qquad (i = 1, \cdots, n);$$

$$\sum_{i=1}^{n} x_{ij} = 1, \qquad (j = 1, \cdots, n).$$

The null space of the associated system of homogeneous equations (see **I.3.1.3, I.3.1.4**) has dimension $n^2 - (2n - 1) = (n - 1)^2$. In order to see this, first observe that the conditions for X to be in D_n are equivalent to the equations (see **1.2.1**)

(i) $XJ = J; JX = J.$

(ii)   The associated homogeneous system becomes $XJ = 0_{n,n}; JX = 0_{n,n}.$

The matrix J is symmetric and has the single nonzero characteristic root equal to n. Hence there exists an orthogonal $S \in M_n(\mathbf{R})$ such that $J = nSE_{1,1}S^{-1}$ (see **I.2.20.9** and **I.4.10.5**). Then the system (ii) is equivalent to

(iii) $YE_{1,1} = 0_{n,n}; E_{1,1}Y = 0_{n,n}$, where $Y = S^{-1}XS$.

The system (iii) states that $Y^{(1)} = 0_n$ and $Y_{(1)} = 0_n$. It follows that the dimension of the null-space of (ii) is precisely $n^2 - (2n - 1) = (n - 1)^2$. Hence if Ω_n contains a p-simplex with vertices $Q_0, \cdots, Q_p$, then $Q_j - Q_0$, $(j = 1, \cdots, p)$, satisfy the system (i) and are linearly independent. Thus $p \leq (n - 1)^2$. We exhibit now an $(n - 1)^2$-simplex contained in Ω_n and thus show that the dimension of Ω_n is $(n - 1)^2$.

Let

$$M = I_{n-2} \dotplus \begin{pmatrix} \frac{1}{2} & \frac{1}{2} \\ \frac{1}{2} & \frac{1}{2} \end{pmatrix}.$$

Consider the polyhedron spanned by the $(n - 1)^2 + 1$ doubly stochastic matrices:

$$I_n, E_{(i),(n-1)}ME^{(j),(n-1)}, \qquad (i, j = 1, \cdots, n - 1) \tag{1}$$

(see **I.3.4, I.3.10**). We show that this polyhedron is an $(n - 1)^2$-simplex. We have to prove that the matrices

$$F_{(i,j)} = E_{(i),(n-1)}ME^{(j),(n-1)} - I_n, \qquad (i, j = 1, \cdots, n - 1),$$

are linearly independent.

Suppose

$$\sum_{i,j=1}^{n-1} \sigma_{ij}F_{(i,j)} = 0.$$

The only $F_{(i,j)}$ with a nonzero entry in the (p,q) position,

$$(p, q = 1, \cdots, n - 2; p \neq q),$$

is $F_{(p,q)}$. Hence $\sigma_{ij} = 0$ for $i, j = 1, \cdots, n - 2$, $(i \neq j)$. Of the remaining matrices only $F_{(i,i)}$ and $F_{(i,n-1)}$ have nonzero entries in the (i,n) position, $(i = 1, \cdots, n - 2)$; all these entries are $\frac{1}{2}$. We must therefore have $\sigma_{ii} + \sigma_{i,n-1} = 0$, $(i = 1, \cdots, n - 2)$.   Similarly, $\sigma_{ii} + \sigma_{n-1,i} = 0$,

$(i = 1, \cdots, n - 2)$. Moreover, by considering the $(i, n - 1)$ position we also get $\frac{1}{2}\sigma_{i,n-1} + \sigma_{n-1,i} = 0$, $(i = 1, \cdots, n - 2)$. It follows that $\sigma_{ii} = \sigma_{i,n-1} = \sigma_{n-1,i} = 0$, $(i = 1, \cdots, n - 2)$. The remaining coefficient, $\sigma_{n-1,n-1}$ must also be equal to 0. Thus $F_{(i,j)}$, $(i, j = 1, \cdots, n - 1)$, are linearly independent and the matrices **1.11**(1) span an $(n - 1)^2$-simplex.

1.12   Linear functionals

Let f be a *linear functional* on a convex polyhedron S; i.e., a real-valued function, defined for every $x \in S$, such that $f(\alpha_1 x_1 + \alpha_2 x_2) = \alpha_1 f(x_1) + \alpha_2 f(x_2)$ for every $x_1, x_2 \in S$ and $\alpha_1, \alpha_2 \in \mathbf{R}$. Then there exist vertices x_m, $x_M \in S$ such that $f(x_m) \leq f(x)$ and $f(x_M) \geq f(x)$ for all $x \in S$. In other words, a linear functional on a convex polyhedron takes its maximum (or its minimum) on a vertex. For, if $x_1, \cdots, x_p$ are the vertices of S and $x = \sum_{j=1}^{p} \theta_j x_j$, $(\theta_j \geq 0; j = 1, \cdots, p;$ and $\sum_{j=1}^{p} \theta_j = 1)$, then $f(x) = f\left(\sum_{j=1}^{p} \theta_j x_j\right) = \sum_{j=1}^{p} \theta_j f(x_j) \leq \max_j f(x_j)$; similarly, $f(x) \geq \min_j f(x_j)$. A more general form of this important theorem will be given in **2.3.9**.

1.13   Example

Let $Z = (z_{ij})$ be a fixed real n-square matrix and let $X = (x_{ij})$ and $f(X) = \sum_{i,j=1}^{n} x_{ij} z_{ij}$. Then f is a linear functional on the convex polyhedron Ω_n and takes its maximum (or its minimum) on a permutation matrix. Thus

$$\max_{X \in \Omega_n} f(X) = \max_{\sigma \in S_n} \sum_{k=1}^{n} z_{i,\sigma(i)}$$

and

$$\min_{X \in \Omega_n} f(X) = \min_{\sigma \in S_n} \sum_{i=1}^{n} z_{i,\sigma(i)}.$$

2   Convex Functions

2.1   Definitions

We shall be concerned here with real valued functions defined on convex subsets of $\mathbf{U}$. Such a function f defined on a convex subset $\mathbf{S}$ of $\mathbf{U}$ is *convex*, if

$$f(\theta x + (1 - \theta)y) \leq \theta f(x) + (1 - \theta)f(y) \tag{1}$$

for any x and y in S and all θ, $(0 \leq \theta \leq 1)$. If the inequality in **2.1(1)**
is strict for all distinct x and y in S and for all θ, $(0 < \theta < 1)$, then f is
strictly convex. A function g is *concave* (*strictly concave*), if $-g$ is convex
(strictly convex). We will also have occasion to consider a convex function
f on $\mathbf{C}^n$ with values in $\mathbf{R}$. This is defined by the inequality **2.1(1)** in which x
and y are now taken to be in some convex subset S of $\mathbf{C}^n$ which contains
$\theta x + (1 - \theta)y$ whenever $x, y \in S$ and $0 \leq \theta \leq 1$.

2.2 Examples

2.2.1 A linear functional on a convex subset S of U is both convex and
concave.

2.2.2 The function $f(X) = \text{tr}\,(X)$ is convex (and concave) on $M_n(\mathbf{R})$.

2.2.3 The function $f(t) = |t|$ is convex on $\mathbf{R}$. For

$$\begin{aligned}
f(\theta t_1 + (1 - \theta)t_2) &= |\theta t_1 + (1 - \theta)t_2| \\
&\leq \theta|t_1| + (1 - \theta)|t_2| \\
&= \theta f(t_1) + (1 - \theta)f(t_2).
\end{aligned}$$

2.3 Properties of convex functions

The following results, with the exception of **2.3.7**, are immediate consequences of the definitions.

2.3.1 If $f_1, \cdots, f_n$ are convex functions on $\mathbf{R}$ and $\sigma_i \geq 0$, $(i = 1, \cdots, n)$,
then $f(x) = \sum\limits_{i=1}^{n} \sigma_i f_i(x)$ is a convex function on $\mathbf{R}$.

2.3.2 If f is a convex function on U, $x_i \in U$, $(i = 1, \cdots, n)$, $\sigma_i \geq 0$,
$(i = 1, \cdots, n)$, and $\sum\limits_{i=1}^{n} \sigma_i = 1$, then $f\left(\sum\limits_{i=1}^{n} \sigma_i x_i\right) \leq \sum\limits_{i=1}^{n} \sigma_i f(x_i)$.

2.3.3 If f is a convex function on U satisfying $a \leq f(x) \leq b$ for all $x \in U$,
and h is an increasing convex function defined on the interval $a \leq t \leq b$,
then the function g defined by $g(x) = h(f(x))$ is convex on U.

2.3.4 If $f_1, \cdots, f_n$ are convex functions on U and f is the function on U
defined by $f(x) = \max\limits_{i=1,\cdots,n} f_i(x)$, then f is convex on U.

2.3.5 If f is a convex function on U and $r \in \mathbf{R}$, then the subset S of U
of all x for which $f(x) \leq r$, $(f(x) < r)$, is convex.

2.3.6 If f_i is convex on an interval F_i in $\mathbf{R}$, $(i = 1, \cdots, n)$, then

$f(x) = \sum\limits_{i=1}^{n} f_i(x_i)$ is convex on the subset **U** of $\mathbf{R}^n$ of all points $x = (x_1, \cdots, x_n)$, $x_i \in F_i$, $(i = 1, \cdots, n)$.

2.3.7   If f is convex on all of **U** and satisfies $f(x) > 0$ for all nonzero x and if there exists $p > 1$ such that $f(\theta x) = \theta^p f(x)$ for all $x \in \mathbf{U}$ and all $\theta \geq 0$ (i.e., f is *homogeneous of degree p*), then $h(x) = (f(x))^{1/p}$ is also convex on all of **U**. (It is assumed that $\theta x \in \mathbf{U}$ whenever $x \in \mathbf{U}$ and $\theta \geq 0$.) Moreover, if the inequality **2.1(1)** for f is strict whenever $0 < \theta < 1$ and neither x nor y is a positive multiple of the other, then under the same conditions the inequality **2.1(1)** with f replaced by h is strict.

To see this, first note that if x is zero, then $f(x) = 0 = h(x)$; otherwise $h(x) > 0$. It follows that if either of x or y is zero, then $h(\theta x + (1 - \theta)y) = \theta h(x) + (1 - \theta)h(y)$. Thus for x and y not zero

$$f\left(\frac{x}{h(x)}\right) = \left(\frac{1}{h(x)}\right)^p f(x) = \frac{f(x)}{f(x)} = 1 = f\left(\frac{y}{h(y)}\right),$$

$$
\begin{aligned}
h\left(\theta \frac{x}{h(x)} + (1 - \theta)\frac{y}{h(y)}\right) &= \left(f\left(\theta \frac{x}{h(x)} + (1 - \theta)\frac{y}{h(y)}\right)\right)^{1/p} \\
&\leq \left(\theta f\left(\frac{x}{h(x)}\right) + (1 - \theta)f\left(\frac{y}{h(y)}\right)\right)^{1/p} \quad (1) \\
&= (\theta + (1 - \theta))^{1/p} \\
&= 1,
\end{aligned}
$$

for all $0 \leq \theta \leq 1$. Then let

$$\theta = \frac{h(x)}{h(x) + h(y)}$$

and observe that **2.3(1)** becomes

$$h\left(\frac{x + y}{h(x) + h(y)}\right) \leq 1, \qquad (2)$$

$$h(x + y) \leq h(x) + h(y).$$

The inequality **2.3(2)** together with the fact that $h(\theta x) = \theta h(x)$ proves the convexity of h. The second statement in **2.3.7** follows immediately.

2.3.8   If f is a function on an interval F of $\mathbf{R}$ and $f''(x)$ exists in the interior of F, then f is convex on F if and only if $f''(x) \geq 0$. If $f''(x) > 0$ in the interior of F, then f is strictly convex on F.

2.3.9   If f is a convex function defined on the polyhedron $H(x_0 \cdots, x_p) \subset \mathbf{U}$, then for all $x \in H(x_0, \cdots, x_p)$

$$f(x) \leq \max_{i=0,\cdots,p} f(x_i). \qquad (3)$$

If f is a concave function defined on the polyhedron $H(x_0, \cdots, x_p) \subset \mathbf{U}$, then for all $x \in H(x_0, \cdots, x_p)$

$$f(x) \geq \min_{i=0,\cdots,p} f(x_i). \tag{4}$$

2.3.10 Let **S** be a convex subset of $\mathbf{R}^n$ such that for any permutation matrix $P \in M_n(\mathbf{R})$, $Pu \in \mathbf{S}$ whenever $u \in \mathbf{S}$. A function f on **S** is *symmetric*, if $f(Pu) = f(u)$ for all $u \in \mathbf{S}$ and all permutation matrices $P \in M_n(\mathbf{R})$. If f is a convex symmetric function on **S** and $u = (u_1, \cdots, u_n) \in \mathbf{R}^n$, then

$$f\left(\frac{\sum_{i=1}^{n} u_i}{n}, \ldots, \frac{\sum_{i=1}^{n} u_i}{n}\right) \leq f(u). \tag{5}$$

To see this, observe that $\sum_{i=1}^{n} u_i/n$ is the value of every coordinate of $\sum_{k=1}^{n} P^k u/n$ where P is the full cycle permutation matrix (see **I.2.16.2**). Then

$$f\left(\frac{\sum_{i=1}^{n} u_i}{n}, \ldots, \frac{\sum_{i=1}^{n} u_i}{n}\right) = f\left(\frac{\sum_{k=1}^{n} P^k u}{n}\right)$$

$$\leq \frac{\sum_{k=1}^{n} f(P^k u)}{n}$$

$$= \frac{\sum_{k=1}^{n} f(u)}{n}$$

$$= f(u),$$

the next but last equality following from the symmetry of f.

It can be proved similarly that if f is a concave symmetric function, then

$$f\left(\frac{\sum_{i=1}^{n} u_i}{n}, \ldots, \frac{\sum_{i=1}^{n} u_i}{n}\right) \geq f(u). \tag{6}$$

2.4 Examples

2.4.1 The function

$$f(t) = \begin{cases} t \log t & \text{if } t > 0, \\ 0 & \text{if } t = 0 \end{cases}$$

is strictly convex on the interval $0 \leq t < \infty$. This is seen by direct application of **2.3.8**.

2.4.2 The function $F(X) = \sum_{i,j=1}^{n} f(x_{ij})$, $(X = (x_{ij}) \in \Omega_n)$, where f is the function in **2.4.1**, is strictly convex on Ω_n. This follows immediately from **2.3.6** and **2.4.1**.

2.4.3   The unit ball B_n in $\mathbf{R}^n$, is a convex set. For, B_n is by definition the set of all $x \in \mathbf{R}^n$ for which $h(x) = \left(\sum\limits_{i=1}^{n} x_i^2 \right)^{1/2} \leq 1$. Observe that the function $f_i(t) = t^2$ is convex on $\mathbf{R}$ by **2.3.8** and hence, by **2.3.6**, $f(x) = \sum\limits_{i=1}^{n} x_i^2$ is convex on $\mathbf{R}^n$. From **2.3.7**, it follows that $(f(x))^{1/2} = h(x)$ is convex and **2.3.5** completes the argument. The convexity of B_n is of course equivalent to the triangle inequality in $\mathbf{R}^n$ (see **I.4.5.4**).

2.4.4   The unit n-cube K_n, defined as the set of all $x = (x_1, \cdots, x_n) \in \mathbf{R}^n$ for which $|x_i| \leq 1$, $(i = 1, \cdots, n)$, is convex. To see this, observe that the function $f_i(x) = |x_i| = \max \{x_i, -x_i\}$ is convex by **2.3.4**. Then K_n is defined by $f(x) \leq 1$ where $f(x) = \max\limits_{i} f_i(x)$. An application of **2.3.5** proves the assertion.

2.4.5   The function $f(x) = \left(\sum\limits_{i=1}^{n} |x_i|^p \right)^{1/p}$, $(p > 1)$, is a convex function on all of $\mathbf{R}^n$. For, by **2.3.4**, $|x_i|$ is convex and, since $p > 1$, **2.3.8** and **2.3.3** yield that $|x_i|^p$ is convex. Hence $\sum\limits_{i=1}^{n} |x_i|^p$ is convex on $\mathbf{R}^n$ by **2.3.6**. It follows from a direct application of **2.3.7** that $f(x)$ is convex.

2.4.6   Let f denote the function in **2.4.5**. Then clearly f is symmetric on all of $\mathbf{R}^n$ and it follows from **2.3.10** that $\left| \sum\limits_{i=1}^{n} u_i \right|^p \leq n^{p-1} \sum\limits_{i=1}^{n} |u_i|^p$.

3   Classical Inequalities

3.1   Power means

3.1.1   If $a_1, \cdots, a_n$ are positive real numbers, then their *power mean of order* r, $\mathfrak{M}_r$, is defined:

$$\mathfrak{M}_r = \begin{cases} \left(\prod\limits_{i=1}^{n} a_i \right)^{1/n} & \text{if } r = 0, \\[2ex] \left(\dfrac{\sum\limits_{i=1}^{n} a_i^r}{n} \right)^{1/r} & \text{if } r \neq 0. \end{cases} \tag{1}$$

The basic result concerning power means follows. For fixed $a_1, \cdots, a_n$ and $r < s$,

$$\mathfrak{M}_r \leq \mathfrak{M}_s \tag{2}$$

with equality, if and only if $a_1 = \cdots = a_n$.

3.1.2 The following inequalities which are special cases of **3.1(2)** will be used extensively:

(arithmetic-geometric mean inequality)

$$\left(\prod_{i=1}^{n} a_i \right)^{1/n} \le \frac{1}{n} \sum_{i=1}^{n} a_i \tag{3}$$

$$\left(\sum_{i=1}^{n} a_i \right)^2 \le n \sum_{i=1}^{n} a_i^2. \tag{4}$$

3.1.3 Let $1 \le p \le 2$ and $a_i, b_i > 0$, $(i = 1, 2, \cdots, n)$; then

$$\frac{\sum_{i=1}^{n} (a_i + b_i)^p}{\sum_{i=1}^{n} (a_i + b_i)^{p-1}} \le \frac{\sum_{i=1}^{n} a_i^p}{\sum_{i=1}^{n} a_i^{p-1}} + \frac{\sum_{i=1}^{n} b_i^p}{\sum_{i=1}^{n} b_i^{p-1}};$$

i.e., the function

$$B_p = \frac{\mathfrak{M}_p^p}{\mathfrak{M}_{p-1}^{p-1}}$$

is convex for $1 \le p \le 2$ on the positive orthant of $\mathbf{R}^n$.

3.2 Symmetric Functions

3.2.1 (Newton's inequalities) Let $a_i \in \mathbf{R}$, $(i = 1, \cdots, n)$, and let $E_r(a_1, \cdots, a_n)$ be the rth elementary symmetric function of the a_i [see **I.2.1(1)**]. Let $p_r = E_r / \binom{n}{r}$ be the rth *weighted elementary symmetric function* of the a_i. Then for fixed $a_1, \cdots, a_n$ and $1 \le r < n$

$$p_{r-1} p_{r+1} \le p_r^2, \tag{1}$$

with equality if and only if the a_i are equal.

3.2.2 In the remaining part of this section $a_1, \cdots, a_{n+1}$ will denote positive real numbers. If $p_1, \cdots, p_n$ are the weighted elementary symmetric functions of $a_1, \cdots, a_n$, then

$$p_1 > p_2^{1/2} > p_3^{1/3} > \cdots > p_n^{1/n} \tag{2}$$

with equality if and only if the a_i are equal.

3.2.3 Let $a = (a_1, \cdots, a_n)$ and let $\bar{a}$ denote the $(n+1)$-tuple $(a_1, \cdots, a_n, a_{n+1})$. If $1 \le r \le k \le n$, then

$$\frac{p_r^k(a)}{p_k^r(a)} \le \frac{p_r^{k+1}(\bar{a})}{p_{k+1}^r(\bar{a})} \tag{3}$$

with equality if and only if $a_1 = \cdots = a_{n+1}$.

3.2.4 Let $A_n = \sum_{i=1}^{n} a_i/n$ and $G_n = \left(\sum_{i=1}^{n} a_i\right)^{1/n}$. Then for $r = 1$, $k = n$, 3.2(3) becomes

$$\left(\frac{A_n}{G_n}\right)^n \leq \left(\frac{A_{n+1}}{G_{n+1}}\right)^{n+1}$$

with equality if and only if $a_1 = \cdots = a_{n+1}$.

3.2.5 Let $a = (a_1, \cdots, a_n)$ and $b = (b_1, \cdots, b_n)$ be n-tuples of positive numbers and $E_r(a)$ denote the rth elementary symmetric function of $a_1, \cdots, a_n$. Then

$$[E_r(a + b)]^{1/r} \geq [E_r(a)]^{1/r} + [E_r(b)]^{1/r},$$

i.e., the function $E_r^{1/r}$ is concave on the positive orthant of $\mathbf{R}^n$, with equality if and only if $a = tb$, $(t > 0)$.

3.2.6 If $a = (a_1, \cdots, a_n)$ and $b = (b_1, \cdots, b_n)$ are n-tuples of nonnegative numbers each with at least r positive coordinates and $1 \leq p \leq r \leq n$, then

$$\left(\frac{E_r(a + b)}{E_{r-p}(a + b)}\right)^{1/p} \geq \left(\frac{E_r(a)}{E_{r-p}(a)}\right)^{1/p} + \left(\frac{E_r(b)}{E_{r-p}(b)}\right)^{1/p},$$

with equality if and only if $a = tb$, $(t > 0)$; i.e., the function $(E_r/E_{r-p})^{1/p}$ is concave on its domain in the nonnegative orthant.

3.2.7 Let $f(a) = T_{(r,k,n)}^{1/r}(a)$ in which

$$T_{(r,k,n)}(a) = \sum_{i_1 + \cdots + i_n = r} \delta_{i_1} \delta_{i_2} \cdots \delta_{i_n} a_1^{i_1} \cdots a_n^{i_n}, \quad (i_j \geq 0; j = 1, \cdots, n),$$

$$\delta_i = \begin{cases} \binom{k}{i} & \text{if } k > 0, \\ (-1)^i \binom{k}{i} & \text{if } k < 0, \end{cases}$$

and k is any number, provided that if k is positive and not an integer, then $n < k + 1$. Then on the nonnegative orthant in $\mathbf{R}^n$ the function f is convex for $k < 0$ and concave for $k > 0$. For $k = 1$, $f(a) = E_r(a)$ and for $k = -1$, $f(a) = h_r(a)$ (see **I.2.15.14**).

3.2.8 Let $h_r(a) = \sum_{\omega \in G_{r,n}} a_\omega$ be the rth completely symmetric function (see **I.2.15.14**) of the n-tuple $a = (a_1, \cdots, a_n)$ of positive numbers. It follows from **3.2.7** that

$$(h_r(a + b))^{1/r} \leq (h_r(a))^{1/r} + (h_r(b))^{1/r},$$

i.e., the function $h_r^{1/r}$ is convex on the positive orthant.

3.3   Hölder inequality

If $x = (x_1, \cdots, x_n)$, $y = (y_1, \cdots, y_n)$ are vectors in the nonnegative orthant of $\mathbf{R}^n$ and $0 < \theta < 1$, then

$$\sum_{j=1}^{n} x_j y_j \leq \left(\sum_{j=1}^{n} x_j^{1/\theta} \right)^{\theta} \left(\sum_{j=1}^{n} y_j^{1/1-\theta} \right)^{1-\theta} \tag{1}$$

with equality, if and only if $(x_1^{1/\theta}, \cdots, x_n^{1/\theta})$ and $(y_1^{1/(1-\theta)}, \cdots, y_n^{1/(1-\theta)})$ are linearly dependent.

Let $\theta_i \geq 0$, $a_i > 0$, $(i = 1, \cdots, k)$, and $\sum_{i=1}^{k} \theta_i = 1$. Since the log function is concave in $(0, \infty)$, we have

$$\log \left(\sum_{i=1}^{k} \theta_i a_i \right) \geq \sum_{i=1}^{k} \theta_i \log a_i = \log \left(\prod_{i=1}^{k} a_i^{\theta_i} \right).$$

Thus

$$\prod_{i=1}^{k} a_i^{\theta_i} \leq \sum_{i=1}^{k} \theta_i a_i. \tag{2}$$

In particular, for any $a, b > 0$ and $0 \leq \theta \leq 1$

$$a^{\theta} b^{1-\theta} \leq \theta a + (1 - \theta) b \tag{3}$$

with equality if and only if either $a = b$ or $\theta = 0$ or 1.

Let $\theta > 0$ and in 3.3(2) set

$$a = \frac{x_j^{1/\theta}}{\sum_{j=1}^{n} x_j^{1/\theta}}, \qquad b = \frac{y_j^{1/(1-\theta)}}{\sum_{j=1}^{n} y_j^{1/(1-\theta)}}.$$

Sum up the resulting inequalities for $j = 1, \cdots, n$:

$$\sum_{j=1}^{n} \left(\frac{x_j^{1/\theta}}{\sum_{j=1}^{n} x_j^{1/\theta}} \right)^{\theta} \left(\frac{y_j^{1/(1-\theta)}}{\sum_{j=1}^{n} y_j^{1/(1-\theta)}} \right)^{1-\theta}$$

$$\leq \sum_{j=1}^{n} \left(\theta \frac{x_j^{1/\theta}}{\sum_{j=1}^{n} x_j^{1/\theta}} + (1 - \theta) \frac{y_j^{1/(1-\theta)}}{\sum_{j=1}^{n} y_j^{1/(1-\theta)}} \right) = 1.$$

This gives the Hölder inequality 3.3(1). The inequality clearly holds, if some (or all) of the x_j, y_j are 0.

3.3.1   The inequality 3.3(1) can be extended to complex numbers:

$$\left| \sum_{j=1}^{n} x_j y_j \right| \leq \left(\sum_{j=1}^{n} |x_j|^{1/\theta} \right)^{\theta} \left(\sum_{j=1}^{n} |y_j|^{1/(1-\theta)} \right)^{1-\theta}. \tag{4}$$

3.3.2 For $\theta = \frac{1}{2}$ the inequality **3.3(1)** becomes

$$\sum_{j=1}^{n} x_j y_j \leq \left(\sum_{j=1}^{n} x_j^2\right)^{1/2} \left(\sum_{j=1}^{n} y_j^2\right)^{1/2}.$$

This is a special case of the Cauchy-Schwarz inequality (see **I.4.5.3**).

3.4 Minkowski inequality

If $a_{ij} \geq 0$, $(i = 1, \cdots, m; j = 1, \cdots, n; 0 < \theta < 1)$, then

$$\left(\sum_{j=1}^{n} \left(\sum_{i=1}^{m} a_{ij}\right)^{1/\theta}\right)^{\theta} \leq \sum_{i=1}^{m} \left(\sum_{j=1}^{n} a_{ij}^{1/\theta}\right)^{\theta} \tag{1}$$

with equality, if and only if the rank of the matrix (a_{ij}) does not exceed 1.

We apply the Hölder inequality to prove **3.4(1)**, the Minkowski inequality. In **3.3(1)** put $x_j = a_{ij}, y_j = \left(\sum_{i=1}^{m} a_{ij}\right)^{(1-\theta)/\theta}$, and observe that

$$\sum_{j=1}^{n} a_{ij} \left(\sum_{i=1}^{m} a_{ij}\right)^{(1-\theta)/\theta} \leq \left(\sum_{j=1}^{n} a_{ij}^{1/\theta}\right)^{\theta} \left(\sum_{j=1}^{n} \left(\sum_{i=1}^{m} a_{ij}\right)^{1/\theta}\right)^{1-\theta}, \quad (i = 1, \cdots, m).$$

Now, sum up with respect to i, noting that

$$\sum_{i=1}^{m} \sum_{j=1}^{n} a_{ij} \left(\sum_{i=1}^{m} a_{ij}\right)^{(1-\theta)/\theta} = \sum_{j=1}^{n} \left(\sum_{i=1}^{m} a_{ij}\right)^{1/\theta}.$$

Thus

$$\sum_{j=1}^{n} \left(\sum_{i=1}^{m} a_{ij}\right)^{1/\theta} \leq \sum_{i=1}^{m} \left(\sum_{j=1}^{n} a_{ij}^{1/\theta}\right)^{\theta} \left(\sum_{j=1}^{n} \left(\sum_{i=1}^{m} a_{ij}\right)^{1/\theta}\right)^{1-\theta},$$

that is,

$$\left(\sum_{j=1}^{n} \left(\sum_{i=1}^{m} a_{ij}\right)^{1/\theta}\right)^{\theta} \leq \sum_{i=1}^{m} \left(\sum_{j=1}^{n} a_{ij}^{1/\theta}\right)^{\theta}.$$

3.4.1 For $m = 2$ the Minkowski inequality takes the form:

$$\left(\sum_{j=1}^{n} (x_j + y_j)^{1/\theta}\right)^{\theta} \leq \left(\sum_{j=1}^{n} x_j^{1/\theta}\right)^{\theta} + \left(\sum_{j=1}^{n} y_j^{1/\theta}\right)^{\theta}$$

for any nonnegative x_j, y_j, and $0 < \theta < 1$ (see **2.4.5**).

3.4.2 For $\theta = \frac{1}{2}$ the formula **3.4.1** gives

$$\left(\sum_{j=1}^{n} (x_j + y_j)^2\right)^{1/2} \leq \left(\sum_{j=1}^{n} x_j^2\right)^{1/2} + \left(\sum_{j=1}^{n} y_j^2\right)^{1/2}$$

which is the triangle inequality in $\mathbf{R}^n$ (see **I.4.5.4**).

3.5 Other inequalities

3.5.1   If $x_i, y_i \geq 0$, $(i = 1, \cdots, n)$, then

$$\left(\prod_{i=1}^{n} (x_i + y_i) \right)^{1/n} \geq \left(\prod_{i=1}^{n} x_i \right)^{1/n} + \left(\prod_{i=1}^{n} y_i \right)^{1/n} \tag{1}$$

with equality, if and only if $(x_1, \cdots, x_n)$ and $(y_1, \cdots, y_n)$ are linearly dependent. Thus the geometric mean is a concave function on the non-negative orthant of $\mathbf{R}^n$. For, 3.5(1) holds trivially, if for some i, $x_i + y_i = 0$. Assume therefore that $x_i + y_i > 0$. Then

$$1 = \frac{1}{n} \sum_{i=1}^{n} \frac{x_i}{x_i + y_i} + \frac{1}{n} \sum_{i=1}^{n} \frac{y_i}{x_i + y_i}$$

$$\geq \left(\prod_{i=1}^{n} \frac{x_i}{x_i + y_i} \right)^{1/n} + \left(\prod_{i=1}^{n} \frac{y_i}{x_i + y_i} \right)^{1/n}, \qquad \text{[by 3.1(3)]},$$

$$= \left(\left(\prod_{i=1}^{n} x_i \right)^{1/n} + \left(\prod_{i=1}^{n} y_i \right)^{1/n} \right) \Big/ \left(\prod_{i=1}^{n} (x_i + y_i) \right)^{1/n}.$$

3.5.2 (Kantorovich inequality)   Let f be a convex function on $[\lambda_n, \lambda_1]$, where $\lambda_1 \geq \cdots \geq \lambda_n > 0$. Suppose that $tf(t) \geq 1$ on $[\lambda_n, \lambda_1]$ and let $c > 0$. On the $(n - 1)$-simplex $H(e_1, \cdots, e_n) = S_{n-1}$ define

$$F(a) = \sum_{i=1}^{n} a_i \lambda_i \sum_{i=1}^{n} a_i f(\lambda_i), \qquad (a \in S_{n-1}).$$

Then

$$1 \leq F(a) \leq \max \left\{ c\lambda_1 + \frac{1}{c} f(\lambda_1), \qquad c\lambda_n + \frac{1}{c} f(\lambda_n) \right\}. \tag{2}$$

The choice $f(t) = t^{-1}$ and $c = (\lambda_1 \lambda_n)^{-1/2}$ yields

$$1 \leq \sum a_i \lambda_i \sum a_i \lambda_i^{-1} \leq \frac{1}{4} \left[\left(\frac{\lambda_1}{\lambda_n} \right)^{1/2} + \left(\frac{\lambda_n}{\lambda_1} \right)^{1/2} \right]^2. \tag{3}$$

The right side of **3.5(3)** is achievable with $a_1 = \frac{1}{2}$, $a_n = \frac{1}{2}$, and $a_i = 0$, $(i \neq 1, n)$.

3.5.3 (Lagrange identity)

$$\left(\sum_{i=1}^{n} a_i^2 \right) \left(\sum_{i=1}^{n} b_i^2 \right) - \left(\sum_{i=1}^{n} a_i b_i \right)^2 = \sum_{\substack{i,j=1 \\ i<j}}^{n} (a_i b_j - a_j b_i)^2. \tag{4}$$

The Cauchy-Schwarz inequality for $\mathbf{R}^n$ follows immediately from **3.5(4)**.

3.5.4   If $x = (x_1, \cdots, x_n)$, $y = (y_1, \cdots, y_n)$, and $x_1 \geq x_2 \geq \cdots \geq x_n$, $y_1 \geq y_2 \geq \cdots \geq y_n$,

$$x_1 + x_2 + \cdots + x_k \leq y_1 + y_2 + \cdots + y_k, (k = 1, \cdots, n - 1),$$
$$x_1 + x_2 + \cdots + x_n = y_1 + y_2 + \cdots + y_n,$$

then there exists $S \in \Omega_n$ such that $x = Sy$. In fact S may be chosen to be orthostochastic (see **1.4.3**).

3.6 Example

The vectors $x = (5, 4, 3, 2)$ and $y = (7, 4, 2, 1)$ satisfy the conditions of
3.5.4. We construct an $S \in \Omega_4$ such that $x = Sy$.

Call the number of nonzero coordinates in $y - x$ the *discrepancy* of x
and y; in this example the discrepancy of x and y is 3. Let

$$S_1 = \begin{pmatrix} \theta & 0 & 1 - \theta & 0 \\ 0 & 1 & 0 & 0 \\ 1 - \theta & 0 & \theta & 0 \\ 0 & 0 & 0 & 1 \end{pmatrix}.$$

Then

$$S_1 y = \begin{pmatrix} 5\theta + 2 \\ 4 \\ -5\theta + 7 \\ 1 \end{pmatrix}.$$

Choose θ so that the discrepancy of x and $S_1 y$ is less than that of x and y;
i.e., so that either $5\theta + 2 = 5$ or $-5\theta + 7 = 3$. The theory guarantees
that at least one of these values for θ lies in the interval $[0, 1]$. Let $\theta = \frac{3}{5}$.
Then

$$S_1 = \begin{pmatrix} \frac{3}{5} & 0 & \frac{2}{5} & 0 \\ 0 & 1 & 0 & 0 \\ \frac{2}{5} & 0 & \frac{3}{5} & 0 \\ 0 & 0 & 0 & 1 \end{pmatrix} \in \Omega_4$$

and

$$S_1 y = \begin{pmatrix} 5 \\ 4 \\ 4 \\ 1 \end{pmatrix}$$

which still differs from x in its last two entries. Let

$$S_2 = \begin{pmatrix} 1 & 0 & 0 & 0 \\ 0 & 1 & 0 & 0 \\ 0 & 0 & \alpha & 1 - \alpha \\ 0 & 0 & 1 - \alpha & \alpha \end{pmatrix}.$$

Then

$$S_2 S_1 y = \begin{pmatrix} 5 \\ 4 \\ 3\alpha + 1 \\ -3\alpha + 4 \end{pmatrix}.$$

Choose α so that $3\alpha + 1 = 3$ and $-3\alpha + 4 = 2$. Again the theory guarantees the existence of such α in the interval $[0, 1]$. Thus $\alpha = \frac{2}{3}$ and

$$S_2 = \begin{pmatrix} 1 & 0 & 0 & 0 \\ 0 & 1 & 0 & 0 \\ 0 & 0 & \frac{2}{3} & \frac{1}{3} \\ 0 & 0 & \frac{1}{3} & \frac{2}{3} \end{pmatrix}.$$

The matrix

$$S = S_2 S_1 = \begin{pmatrix} \frac{3}{5} & 0 & \frac{2}{5} & 0 \\ 0 & 1 & 0 & 0 \\ \frac{4}{15} & 0 & \frac{2}{5} & \frac{1}{3} \\ \frac{2}{15} & 0 & \frac{1}{5} & \frac{2}{3} \end{pmatrix}$$

is in Ω_4 and has the property that $Sy = x$.

In general, the method is to construct a sequence $\{S_j\}$ of doubly stochastic matrices with 1's in all but two rows (the hth and the kth row) such that $S_j S_{j-1} \cdots S_1 y$ and x have at least one discrepancy less than $S_{j-1} \cdots S_1 y$ and x; k is the least number such that $(S_{j-1} \cdots S_1 y)_k - x_k < 0$ and h is the largest number less than k and such that $(S_{j-1} \cdots S_1 y)_h - x_h > 0$.

4   Convex Functions and Matrix Inequalities

4.1   Convex functions of matrices

Given a convex function f and a normal matrix A we obtain a result relating the values of f on a set of quadratic forms for A to the values of f on some subset of the characteristic roots of A. Some important theorems which amount to making special choices for A and f will follow in the remaining subsections of **4.1**.

4.1.1   Let f be a function on $\mathbf{C}^k$ with values in $\mathbf{R}$. If $\lambda = (\lambda_1, \cdots, \lambda_n) \in \mathbf{C}^n$ and $S \in \Omega_n$, then we may define a function g on Ω_n to $\mathbf{R}$ by

$$g(S) = f((S_{(1)}, \lambda), \cdots, (S_{(k)}, \lambda)). \tag{1}$$

4.1.2   If the function f in **4.1.1** is convex (concave), then g is a convex (concave) function on Ω_n with values in $\mathbf{R}$. For, if S and T are in Ω_n and $0 \leq \theta \leq 1$, then

$$\begin{aligned} g(\theta S + (1 - \theta)T) &= f(((\theta S + (1 - \theta)T)_{(1)}, \lambda), \cdots, ((\theta S + (1 - \theta)T)_{(k)}, \lambda)) \\ &= f(\theta(S_{(1)}, \lambda) + (1 - \theta)(T_{(1)}, \lambda), \cdots, \theta(S_{(k)}, \lambda) \\ &\qquad + (1 - \theta)(T_{(k)}, \lambda)) \\ &\leq \theta f((S_{(1)}, \lambda), \cdots, (S_{(k)}, \lambda)) \\ &\qquad + (1 - \theta)f((T_{(1)}, \lambda), \cdots, (T_{(k)}, \lambda)) \\ &= \theta g(S) + (1 - \theta)g(T). \end{aligned}$$

4.1.3   Let A be a normal matrix in $M_n(\mathbf{C})$ with characteristic roots $\lambda_1, \cdots, \lambda_n$. Let $x_1, \cdots, x_k$ be an orthonormal set of vectors in $\mathbf{C}^n$. Set

$z_j = (Ax_j, x_j)$, $(j = 1, \cdots, k)$, $z = (z_1, \cdots, z_k)$, $\lambda = (\bar{\lambda}_1, \cdots, \bar{\lambda}_n)$. Then there exists a doubly stochastic matrix $S \in \Omega_n$ such that

$$z_j = (S_{(j)}, \lambda), \quad (j = 1, \cdots, k). \tag{2}$$

The result is easily seen as follows. Since A is normal, there exists (see **I.4.10.3**) an orthonormal basis of characteristic vectors $u_1, \cdots, u_n$ in $\mathbf{C}^n$ such that $Au_j = \lambda_j u_j$, $(j = 1, \cdots, n)$. But then, by **I.4.5.5**,

$$
\begin{aligned}
Ax_j &= A \sum_{t=1}^{n} (x_j, u_t)u_t \\
&= \sum_{t=1}^{n} (x_j, u_t)\lambda_t u_t,
\end{aligned}
$$

and hence from the orthonormality of $u_1, \cdots, u_n$

$$
\begin{aligned}
(Ax_j, x_j) &= \sum_{t=1}^{n} \lambda_t (x_j, u_t)\overline{(x_j, u_t)} \\
&= \sum_{t=1}^{n} \lambda_t |(x_j, u_t)|^2, \qquad (j = 1, \cdots, k). \tag{3}
\end{aligned}
$$

Let $x_{k+1}, \cdots, x_n$ be a completion of $x_1, \cdots, x_k$ to an orthonormal basis of $\mathbf{C}^n$ (see **I.4.5.1**). Let $S \in M_n(\mathbf{R})$ be the matrix whose (j,k) entry is $|(x_j, u_k)|^2$. The orthonormality of the vectors $u_1, \cdots, u_n$ implies that

$$1 = ||x_j||^2 = \sum_{t=1}^{n} |(x_j, u_t)|^2, \qquad (j = 1, \cdots, n),$$

and similarly the orthonormality of the vectors $x_1, \cdots, x_n$ implies that

$$1 = ||u_t||^2 = \sum_{j=1}^{n} |(x_j, u_t)|^2, \qquad (t = 1, \cdots, n).$$

Hence $S \in \Omega_n$ and thus **4.1(3)** becomes

$$z_j = (S_{(j)}, \lambda), \qquad (j = 1, \cdots, k).$$

4.1.4   Let f be a real valued convex function on the convex subset L of $\mathbf{C}^n$ (see **2.1**) where L is defined as the totality of points $S\lambda$, for $S \in \Omega_n$ and $\bar{\lambda} = (\lambda_1, \cdots, \lambda_n)$ the n-tuple of characteristic roots of a normal matrix $A \in M_n(\mathbf{C})$. Then

$$g(S) = f((S_{(1)}, \lambda), \cdots, (S_{(k)}, \lambda))$$

is convex on Ω_n (see **4.1.2**) and, moreover, $g(S) \leq g(P)$ where P is an appropriate permutation matrix in Ω_n. It follows that if f is convex on L, then

$$\max_{(x_i, x_j) = \delta_{ij}} f((Ax_1, x_1), \cdots, (Ax_k, x_k)) = f(\lambda_{i_1}, \cdots, \lambda_{i_k}) \tag{4}$$

where $(i_1, \cdots, i_k) \in D_{k,n}$ (see **I.2.1**); i.e., $i_1, \cdots, i_k$ are some k distinct integers chosen from $1, \cdots, n$. The notation simply means that we maximize the left side of **4.1(4)** over all orthonormal sets $x_1, \cdots, x_k$ in $\mathbf{C}^n$.

If f is concave on L, then

$$\min_{(x_i, x_j) = \delta_{ij}} f((Ax_1, x_1), \cdots, (Ax_k, x_k)) = f(\lambda_{j_1}, \cdots, \lambda_{j_k}) \tag{5}$$

where $(j_1, \cdots, j_k) \in D_{k,n}$. This result follows immediately from **4.1.1**, **4.1.2**, **4.1.3**, **2.3.9**, and **1.7**.

4.1.5 If A is a hermitian matrix in $M_n(\mathbf{C})$ and $x_1, \cdots, x_k$ are orthonormal vectors in $\mathbf{C}^n$, then

$$\sum_{j=1}^{k} \lambda_{n-j+1} \le \sum_{j=1}^{k} (Ax_j, x_j) \le \sum_{j=1}^{k} \lambda_j$$

where $\lambda_1 \ge \cdots \ge \lambda_n$ are the characteristic roots of A. This follows from **4.1.4** by choosing

$$f(t_1, \cdots, t_k) = \sum_{j=1}^{k} t_j.$$

4.1.6 If A is a nonnegative hermitian matrix in $M_n(\mathbf{C})$ with characteristic roots $\lambda_1 \ge \cdots \ge \lambda_n \ge 0$ and $x_1, \cdots, x_k$ are orthonormal vectors in $\mathbf{C}^n$, then

$$\prod_{j=1}^{k} \lambda_{n-j+1} \le \prod_{j=1}^{k} (Ax_j, x_j) \le \left(\frac{\sum_{j=1}^{k} \lambda_j}{k} \right)^k.$$

The upper bound follows from **3.1(3)** and **4.1.5**. The lower bound follows from **3.5.1** by choosing

$$f(t_1, \cdots, t_k) = \left(\prod_{j=1}^{k} t_j \right)^{1/k}.$$

4.1.7 **(Hadamard determinant theorem)** If $A \in M_n(\mathbf{C})$, then

$$|d(A)|^2 \le \prod_{j=1}^{n} (A^{(j)}, A^{(j)}) = \prod_{j=1}^{n} \sum_{i=1}^{n} |a_{ij}|^2$$

with equality, if and only if A^*A is a diagonal matrix or A has a zero column. For, $|d(A)|^2 = d(A^*A)$ and A^*A is nonnegative hermitian for any $A \in M_n(\mathbf{C})$ (see **I.4.12.2**). Taking $k = n$ in the lower bound in **4.1.6**, replacing A by A^*A, and choosing $x_j = e_j$, $(j = 1, \cdots, n)$, (see **I.2.19**) we have

$$|d(A)|^2 = d(A^*A) \le \prod_{j=1}^{n} (A^*Ae_j, e_j)$$

$$= \prod_{j=1}^{n} (A^{(j)}, A^{(j)})$$

$$= \prod_{j=1}^{n} \sum_{i=1}^{n} |a_{ij}|^2.$$

In case A is nonnegative hermitian, the following statement holds:

$$d(A) \le \prod_{j=1}^{n} a_{jj}$$

with equality, if and only if A is a diagonal matrix, or A has a zero row and column.

It has recently been proved by one of the authors that there exists an analogue of the Hadamard theorem for permanents.

If $A \in M_n(\mathbf{C})$ and A is nonnegative hermitian, then $p(A) \geq \prod\limits_{i=1}^{n} a_{ii}$ with equality holding if and only if A is a diagonal matrix, or A has a zero row.

4.1.8 (Minkowski determinant theorem) If A and B are nonnegative hermitian matrices in $M_n(\mathbf{C})$, then

$$(d(A + B))^{1/n} \geq (d(A))^{1/n} + (d(B))^{1/n}. \tag{6}$$

This inequality is equivalent to the statement that the function $d^{1/n}$ is concave on the set of n-square nonnegative hermitian matrices in $M_n(\mathbf{C})$. Note that if $0 \leq \theta \leq 1$ and A and B are nonnegative hermitian then $\theta A + (1 - \theta)B$ is nonnegative hermitian and from **4.1(6)** it follows that

$$\begin{aligned} (d(\theta A + (1 - \theta)B))^{1/n} &\geq (d(\theta A))^{1/n} + (d((1 - \theta)B))^{1/n} \\ &= \theta (d(A))^{1/n} + (1 - \theta)(d(B))^{1/n}. \end{aligned} \tag{7}$$

From **4.1(7)** and **3.3(3)** we have

$$d(\theta A + (1 - \theta)B) \geq (d(A))^{\theta}(d(B))^{1-\theta}. \tag{8}$$

To prove **4.1(6)** let $x_1, \cdots, x_n$ be an orthonormal set of characteristic vectors for $(A + B)$ so that

$$\begin{aligned} (d(A + B))^{1/n} &= \left(\prod_{i=1}^{n} ((A + B)x_i, x_i) \right)^{1/n} \\ &= \left(\prod_{i=1}^{n} ((Ax_i, x_i) + (Bx_i, x_i)) \right)^{1/n} \\ &\geq \left(\prod_{i=1}^{n} (Ax_i, x_i) \right)^{1/n} + \left(\prod_{i=1}^{n} (Bx_i, x_i) \right)^{1/n} \\ &\geq (d(A))^{1/n} + (d(B))^{1/n}. \end{aligned}$$

The next but last inequality follows from **3.5.1** and the last inequality is just the lower bound in **4.1.6** with $k = n$.

4.1.9 If $A \in M_n(\mathbf{C})$ has characteristic roots $\lambda_1, \cdots, \lambda_n, |\lambda_1| \geq \cdots \geq |\lambda_n|$, and singular values $\alpha_1 \geq \cdots \geq \alpha_n$ (see **I.4.12.2**), then

$$\prod_{j=1}^{k} |\lambda_j| \leq \prod_{j=1}^{k} \alpha_j, \qquad (k = 1, \cdots, n), \tag{9}$$

with equality for $k = n$.

The proof of this statement for $k = 1$ is easy. Simply take x_1 to be a unit characteristic vector of A corresponding to the characteristic root λ_1. Then $|\lambda_1|^2 = (Ax_1, Ax_1) = (A^*Ax_1, x_1)$. But A^*A is nonnegative hermitian and its largest characteristic root is α_1^2. Hence, by **4.1.5** with $k = 1$,

$(A^*Ax_1, x_1) \leq \alpha_1^2$, $|\lambda_1| \leq \alpha_1$. Now apply the preceding to the matrix $C_k(A)$. The largest (in modulus) characteristic root of $C_k(A)$ is $\prod\limits_{j=1}^{k} \lambda_j$ (see **I.2.15.12**). Observe that (see **I.2.7.1**, **I.2.7.2**) $(C_k(A))^*C_k(A) = C_k(A^*)C_k(A) = C_k(A^*A)$ and so the largest singular value of $C_k(A)$ is $\prod\limits_{j=1}^{k} \alpha_j$. Thus the result for $k = 1$, applied to $C_k(A)$, yields **4.1(9)**. The equality in **4.1(9)** occurs for $k = n$ because $|d(A)| = (d(A^*A))^{1/2}$.

4.1.10 If A and B are nonnegative hermitian matrices, define $E_p(A)$, $h_p(A)$, $B_p(A)$ to be respectively the elementary symmetric function of the characteristic roots of A [see **I.2.1(1)**], the completely symmetric function of the characteristic roots of A (see **I.2.15.14**) and finally the function defined in **3.1.3** all evaluated on the characteristic roots of A. Then

$$\left(\frac{E_p(A + B)}{E_{p-k}(A + B)}\right)^{1/k} \geq \left(\frac{E_p(A)}{E_{p-k}(A)}\right)^{1/k} + \left(\frac{E_p(B)}{E_{p-k}(B)}\right)^{1/k}, \tag{10}$$

$$(n \geq p \geq k \geq 1);$$

$$h_p^{1/p}(A + B) \leq h_p^{1/p}(A) + h_p^{1/p}(B), \qquad (p = 1, \cdots, n); \tag{11}$$

$$B_p(A + B) \leq B_p(A) + B_p(B), \qquad (p = 1, \cdots, n). \tag{12}$$

It is assumed in **4.1(10)** that $\rho(A) \geq p - k$ and $\rho(B) \geq p - k$ so that neither denominator is 0. Also $E_0(A)$ is taken to be 1. Note that for $k = p = n$ the inequality **4.1(10)** is just **4.1.8** (Minkowski determinant theorem). These results all follow directly by applying **4.1.4** to the functions in question.

4.2   Inequalities of H. Weyl

If $A \in M_n(\mathbf{C})$ has characteristic roots $\lambda_1, \cdots, \lambda_n$, $|\lambda_1| \geq \cdots \geq |\lambda_n|$, and singular values $\alpha_1 \geq \cdots \geq \alpha_n$, and $s > 0$, then

$$\sum_{j=1}^{k} |\lambda_j|^s \leq \sum_{j=1}^{k} \alpha_j^s, \qquad (k = 1, \cdots, n). \tag{1}$$

The proof of this result runs as follows: We may clearly assume A is nonsingular so that $|\lambda_n| > 0$. Then by taking the logarithm on both sides of **4.1(9)** it follows that

$$\sum_{j=1}^{k} \log |\lambda_j| \leq \sum_{j=1}^{k} \log \alpha_j, \qquad (k = 1, \cdots, n),$$

with equality for $k = n$. Let $u = (\log |\lambda_1|, \cdots, \log |\lambda_n|)$ and $v = (\log \alpha_1, \cdots, \log \alpha_n)$. By **3.5.4**, $u = Sv$ for $S \in \Omega_n$. Next define $f(t) = f(t_1, \cdots, t_n) = \sum\limits_{j=1}^{k} e^{st_j}$. The function f is convex on the positive orthant (see **2.3.6**). Hence

$$g(T) = f((T_{(1)}, v), \cdots, (T_{(n)}, v)) = f(Tv), \qquad (T \in \Omega_n),$$

is convex on Ω_n. From **4.1.4**, $g(T) \leq g(P)$ where P is some permutation matrix. Thus taking $T = S$,

$$f(u) = f(Sv) = g(S) \leq g(P) = f((P_{(1)}, v), \cdots, (P_{(n)}, v)) \qquad (2)$$
$$= f(v_{\sigma(1)}, \cdots, v_{\sigma(n)})$$

for an appropriate $\sigma \in S_n$. Evaluating both sides of the inequality **4.2(2)** gives

$$\sum_{j=1}^{k} e^{s(\log |\lambda_j|)} \leq \sum_{j=1}^{k} e^{s(\log \alpha_{\sigma(j)})}$$

or

$$\sum_{j=1}^{k} |\lambda_j|^s \leq \sum_{j=1}^{k} \alpha_{\sigma(j)}^s \leq \sum_{j=1}^{k} \alpha_j^s.$$

4.3　Kantorovich inequality

4.3.1　If $A \in M_n(\mathbf{C})$ is positive definite hermitian with characteristic roots $\lambda_1 \geq \cdots \geq \lambda_n > 0$ and $x \in \mathbf{C}^n$ is a unit vector, then

$$1 \leq (Ax, x)(A^{-1}x, x) \leq \frac{1}{4}\left(\left(\frac{\lambda_1}{\lambda_n}\right)^{1/2} + \left(\frac{\lambda_n}{\lambda_1}\right)^{1/2}\right)^2.$$

This result follows directly from **3.5(3)**. For if $u_1, \cdots, u_n$ are orthonormal characteristic vectors corresponding to $\lambda_1, \cdots, \lambda_n$ respectively, then

$$(Ax, x) = \sum_{i=1}^{n} a_i \lambda_i, \qquad (A^{-1}x, x) = \sum_{i=1}^{n} a_i \lambda_i^{-1}$$

where $a_i = |(x, u_i)|^2$.

4.3.2　If A is the matrix in **4.3.1** and $\omega \in Q_{k,n}$, then

$$1 \leq d(A[\omega|\omega])d(A^{-1}[\omega|\omega]) \leq \frac{1}{4}\left(\left(\prod_{j=1}^{k} \frac{\lambda_j}{\lambda_{n-j+1}}\right)^{1/2} + \left(\prod_{j=1}^{k} \frac{\lambda_{n-j+1}}{\lambda_j}\right)^{1/2}\right)^2.$$

This is just the inequality in **4.3.1** applied to the matrix $C_k(A)$ (see **I.4.12.7, I.2.15.12**).

4.4　More inequalities

4.4.1　If A and B in $M_n(\mathbf{C})$ are nonnegative hermitian, then

$$d(A + B) \geq d(A) + d(B). \qquad (1)$$

This follows immediately from **4.1.8**.

4.4.2　If $A \in M_n(\mathbf{C})$ is nonnegative hermitian and $\omega \in Q_{k,n}$, then

$$d(A) \leq d(A[\omega|\omega])d(A(\omega|\omega)). \qquad (2)$$

For let P be the diagonal matrix with -1 in the main diagonal positions corresponding to ω and 1 in the remaining main diagonal positions. Then $B = PAP^T$ is also nonnegative hermitian and $B[\omega|\omega] = A[\omega|\omega]$, $B[\omega|\omega) = A(\omega|\omega)$, $B[\omega|\omega) = -A[\omega|\omega)$, and $B(\omega|\omega] = -A(\omega|\omega]$. Then the Laplace expansion theorem (see **I.2.4.11**) and **4.1.8** applied to $A + B$ yield

$$(d(2A[\omega|\omega])d(2A(\omega|\omega)))^{1/n} = (d(A + B))^{1/n}$$
$$\geq (d(A))^{1/n} + (d(B))^{1/n}$$
$$= 2(d(A))^{1/n},$$

and **4.4(2)** follows.

4.4.3 If A and B are in $M_n(\mathbf{C})$, then

$$|p(AB)|^2 \leq p(AA^*)p(B^*B) \tag{3}$$

with equality, if and only if one of the following occurs:

(i)   a row of A or a column of B consists of 0,
(ii) no row of A and no column of B consists of 0 and $A^* = BDP$ where D is a diagonal matrix, and P is a permutation matrix.

In particular,
$$|p(A)|^2 \leq p(AA^*), \qquad |p(A)|^2 \leq p(A^*A).$$

To see **4.4(3)**, apply the Schwarz inequality (see **I.4.5.3**, **I.4.6(2)**, **I.2.12(1)**, **I.2.12.2**) to obtain

$$|p(AB)|^2 = |(P_n(AB)e_1 \cdots e_n, e_1 \cdots e_n)|^2$$
$$= |(P_n(A)P_n(B)e_1 \cdots e_n, e_1 \cdots e_n)|^2$$
$$= |(P_n(B)e_1 \cdots e_n, P_n(A^*)e_1 \cdots e_n)|^2$$
$$\leq (P_n(B)e_1 \cdots e_n, P_n(B)e_1 \cdots e_n)(P_n(A^*)e_1 \cdots e_n, P_n(A^*)e_1 \cdots e_n)$$
$$= (P_n(B^*B)e_1 \cdots e_n, e_1 \cdots e_n)(P_n(AA^*)e_1 \cdots e_n, e_1 \cdots e_n)$$
$$= p(AA^*)p(B^*B).$$

We note that, by **I.4.6(2)**, $e_1 \cdots e_n$ is a unit vector. The discussion of equality is omitted.

4.4.4 If $A \in M_n(\mathbf{C})$ is nonnegative hermitian, then

$$d(A) \leq p(A). \tag{4}$$

For, by **I.5.10.1**, there exists $X \in M_n(\mathbf{C})$ such that X is upper triangular and $X^*X = A$. Hence, by **4.4.3**,

$$d(A) = d(X^*X) = d(X^*)d(X)$$
$$= p(X^*)p(X)$$
$$\leq \sqrt{p(X^*X)p(X^*X)}$$
$$= p(A).$$

4.4.5 If $A \in M_n(\mathbf{C})$ is nonnegative hermitian with characteristic roots $\lambda_1 \geq \cdots \geq \lambda_n \geq 0$ and $\omega \in Q_{k,n}$, then

$$\prod_{j=1}^{k} \lambda_{n-j+1} \leq d(A[\omega|\omega]) \leq \prod_{j=1}^{k} \lambda_j, \tag{5}$$

and

$$d(A[\omega|\omega]) \leq p(A[\omega|\omega]). \tag{6}$$

The inequality 4.4(6) follows from 4.4(4) whereas 4.4(5) results from applying the fact (see **4.1.5**) that $(C_k(A)e_{i_1} \wedge \cdots \wedge e_{i_k}, e_{i_1} \wedge \cdots \wedge e_{i_k})$ is between the largest and smallest characteristic roots of $C_k(A)$ (see **I.2.15.12**). We note that by **I.4.6(1)**, $e_{i_1} \wedge \cdots \wedge e_{i_k}$ is a unit vector.

4.4.6   Let $A \in M_n(\mathbf{C})$ be nonnegative hermitian and let

$$P_k = \prod_{\omega \in Q_{k,n}} d(A[\omega|\omega]).$$

Then

$$P_1 \geq P_2^{1/\binom{n-1}{1}} \geq \cdots \geq P_{n-1}^{1/\binom{n-1}{n-2}} \geq P_n. \tag{7}$$

To prove this observe first that the Hadamard determinant theorem (see **4.1.7**) together with **I.2.7(6)** applied to $C_{n-1}(A)$ implies

$$(d(A))^{n-1} = d(C_{n-1}(A)) \leq \prod_{\omega \in Q_{n-1,n}} d(A[\omega|\omega]). \tag{8}$$

If $\omega \in Q_{k,n}$, there are precisely $n - k$ sequences $\tau \in Q_{k+1,n}$ which contain ω; i.e., $\omega \subset \tau$. Thus using **I.4.12.5** and 4.4(8) on the $(k + 1)$-square principal submatrices of A,

$$P_{k+1} = \prod_{\tau \in Q_{k+1,n}} d(A[\tau|\tau]) \leq \prod_{\tau \in Q_{k+1,n}} \left(\prod_{\omega \in Q_{k,n}, \omega \subset \tau} d(A[\omega|\omega]) \right)^{1/k}$$
$$= \left(\prod_{\omega \in Q_{k,n}} d(A[\omega|\omega]) \right)^{(n-k)/k}$$
$$= P_k^{(n-k)/k},$$

and hence

$$P_{k+1}^{1/\binom{n-1}{k}} \leq P_k^{1/\binom{n-1}{k-1}}.$$

4.4.7   **(Cauchy inequalities)**   If $A \in M_n(\mathbf{C})$ and A is hermitian, $\omega \in Q_{n-r,n}$, then

$$\lambda_{s+r}(A) \leq \lambda_s(A[\omega|\omega]) \leq \lambda_s(A), \qquad (s = 1, \cdots, n - r). \tag{9}$$

(Recall the notational convention of **I.2.15(1)** that the characteristic roots decrease as the subscript increases.)

4.4.8   If A and B are in $M_n(\mathbf{C})$ and are both positive definite hermitian and $\omega = (1, \cdots, k)$, then for each $k = 1, \cdots, n$,

$$\frac{d(A + B)}{d((A + B)[\omega|\omega])} \geq \frac{d(A)}{d(A[\omega|\omega])} + \frac{d(B)}{d(B[\omega|\omega])}.$$

Also,

$$\frac{d(A + B)}{d((A + B)(i|i))} \geq \frac{d(A)}{d(A(i|i))} + \frac{d(B)}{d(B(i|i))}, \qquad (i = 1, \cdots, n).$$

4.4.9   Suppose that A and B in $M_n(\mathbf{C})$ are nonnegative hermitian, $Q \in M_n(\mathbf{C})$, and $A^2 - Q^*Q$ and $B^2 - QQ^*$ are nonnegative hermitian. If $0 \leq p \leq 1$ and $u_1, \cdots, u_k, v_1, \cdots, v_k$ are any $2k$ vectors, then

$$|d((Qu_i, v_j))|^2 \leq \prod_{i=1}^{k} ||A^p u_i||^2 ||B^{1-p} v_i||^2.$$

4.4.10 If A and B are in $M_n(\mathbf{C})$ and $I - A^*A$ and $I - B^*B$ are non-negative hermitian, then

$$|d(I_n - A^*B)|^2 \geq d(I_n - A^*A)d(I_n - B^*B).$$

4.4.11 If $A \in M_n(\mathbf{C})$, $\rho(A) = k$, and $1 \leq r \leq k$, then

$$\sum_{\omega, \tau \in Q_{r,n}} |d(A[\omega|\tau])|^2 \leq \binom{k}{r} k^{-r} (\text{tr } (AA^*))^r,$$

with equality, if and only if $A = V(\alpha I_k \dotplus 0_{n-k,n-k})W$, $(\alpha > 0$, V and W unitary). Note that tr (AA^*) is the square of the Euclidean norm of A (see **I.2.8.2**).

4.4.12 If $A \in M_n(\mathbf{C})$ is nonnegative hermitian, then

$$p(A) \leq \frac{\text{tr } (A^n)}{n}.$$

4.4.13 If $A \in \Omega_n$, then

$$p(A) \leq \left(\frac{\rho(A)}{n}\right)^{1/2}$$

with equality, if and only if A is a permutation matrix.

4.4.14 If A and B in $M_n(\mathbf{C})$ are hermitian and $(i_1, \cdots, i_k) \in Q_{k,n}$, then

$$\sum_{t=1}^{k} \lambda_{i_t}(A + B) \leq \sum_{t=1}^{k} \lambda_{i_t}(A) + \sum_{t=1}^{k} \lambda_t(B).$$

4.4.15 If $A \in M_n(\mathbf{C})$ is hermitian, $N_r = \sum_{\omega \in Q_{r,n}} d(A[\omega|\omega])$, $(N_0 = 1)$, then

$$\rho(A) - r \geq (r + 1)N_{r-1}N_{r+1}(rN_r^2 - (r + 1)N_{r-1}N_{r+1})^{-1}$$

for $1 \leq r \leq \rho(A)$. Equality holds, if and only if $\rho(A) = r$ or $\rho(A) > r$ and all nonzero characteristic roots of A are equal.

4.4.16 If $A \in M_n(\mathbf{C})$ is nonnegative hermitian, $\omega \in Q_{n-k,n}$, then

$$\lambda_{s+t+1}(A) \leq \lambda_{s+1}(A[\omega|\omega]) + \lambda_{t+1}(A(\omega|\omega)),$$
$$(0 \leq s \leq n - k - 1, 0 \leq t \leq k - 1).$$

4.5 Hadamard product

If A and B are in $M_n(\mathbf{C})$, then their *Hadamard product* is the matrix $H = A*B \in M_n(\mathbf{C})$ whose (i,j) entry is $a_{ij}b_{ij}$.

4.5.1 The Hadamard product $A * B$ is a principal n-square submatrix of the Kronecker product $A \otimes B$.

4.5.2 If A and B are nonnegative hermitian matrices, then so is $A * B$ and

$$\lambda_n(A)\lambda_n(B) \leq \lambda_1(A * B) \leq \lambda_1(A)\lambda_1(B). \tag{1}$$

To prove **4.5(1)**, directly apply **4.5.1**, **I.2.15.11**, and **4.4.7**.

4.5.3 If $A \in M_n(\mathbf{C})$ is nonnegative hermitian and $h_{ij} = |a_{ij}|^2$, then $H = (h_{ij})$ is nonnegative hermitian and

$$\lambda_1(H) \leq (\lambda_1(A))^2.$$

For, $H = A * A^T$ is a principal submatrix of $A \otimes A^T$ and is thereby nonnegative hermitian. Apply **4.5.2**.

4.5.4 If A and B in $M_n(\mathbf{C})$ are nonnegative hermitian, then

$$d(A * B) + d(A)d(B) \geq d(A) \prod_{i=1}^{n} b_{ii} + d(B) \prod_{i=1}^{n} a_{ii}.$$

5 Nonnegative Matrices

5.1 Introduction

Most of the inequalities and other results in the preceding sections of this chapter concern positive or nonnegative real numbers. The theorems in section 4 are mostly about positive definite or positive semidefinite matrices which are, in a sense, a matrix generalization of the positive or nonnegative numbers. A more direct matrix analogue of a positive (nonnegative) number is a positive (nonnegative) matrix.

A real n-square matrix $A = (a_{ij})$ is called *positive* (*nonnegative*), if $a_{ij} > 0$ ($a_{ij} \geq 0$) for $i, j = 1, \cdots, n$. We write $A > 0$ ($A \geq 0$). Similarly, an n-tuple $x = (x_1, \cdots, x_n) \in \mathbf{R}^n$ is *positive* (*nonnegative*) if $x_i > 0$ ($x_i \geq 0$), ($i = 1, \cdots, n$). We write $x > 0$, if x is positive and $x \geq 0$, if x is nonnegative.

Doubly stochastic matrices, which played such an important part in the preceding section, form a subset of the set of nonnegative matrices. We shall discuss doubly stochastic matrices in detail in **5.11** and **5.12**.

Some obvious properties of nonnegative n-square matrices can be deduced from the fact that they form a convex set and a partially ordered set. The main interest, however, lies in their remarkable spectral properties. These were discovered by Perron for positive matrices and amplified and generalized for nonnegative matrices by Frobenius. The essence of Frobenius' generalization lies in the observation that a positive square matrix is a special case of a nonnegative indecomposable matrix.

5.2 Indecomposable matrices

A nonnegative n-square matrix $A = (a_{ij})$ $(n > 1)$ is said to be *decomposable* if there exists an ordering $(i_1, \cdots, i_p, j_1, \cdots, j_q)$ of $(1, \cdots, n)$ such that $a_{i_\alpha j_\beta} = 0$, $(\alpha = 1, \cdots, p; \beta = 1, \cdots, q)$. Otherwise A is *indecomposable*.

In other words, A is decomposable, if there exists a permutation matrix P such that $PAP^T = \begin{pmatrix} B & 0 \\ C & D \end{pmatrix}$ where B and D are square. Some writers use the terms *reducible* and *irreducible* instead of decomposable and indecomposable. There is no direct connection between decomposability in the above sense and reducibility as defined in **I.3.31**.

Some preliminary properties of indecomposable n-square matrices follow.

5.2.1 Let $e_1, \cdots, e_n$ be the n-vectors defined in **I.2.19**. Then $A \geq 0$ is decomposable, if and only if there exists a proper subset $\{j_1, \cdots, j_p\}$ of $\{1, \cdots, n\}$ such that
$$\langle Ae_{j_1}, \cdots, Ae_{j_p} \rangle \subset \langle e_{j_1}, \cdots, e_{j_p} \rangle.$$

For example, the full-cycle permutation matrix
$$P = \begin{pmatrix} 0 & 1 & 0 & \cdots & 0 & 0 \\ 0 & 0 & 1 & \cdots & 0 & 0 \\ \cdots & \cdots & \cdots & \cdots & \cdots & \cdots \\ 0 & 0 & 0 & \cdots & 0 & 1 \\ 1 & 0 & 0 & \cdots & 0 & 0 \end{pmatrix} \tag{1}$$

is indecomposable because $Pe_j = e_{j-1}$ (e_0 is to be interpreted as e_n) and therefore $\langle Pe_{j_1}, \cdots, Pe_{j_p} \rangle \subset \langle e_{j_1}, \cdots, e_{j_p} \rangle$ means
$$\langle e_{j_1-1}, \cdots, e_{j_p-1} \rangle = \langle e_{j_1}, \cdots, e_{j_p} \rangle,$$
an impossibility unless $p = n$.

5.2.2 If $A \geq 0$ is indecomposable and $y \geq 0$ is an n-tuple with k zeros, $(1 \leq k < n)$, then the number of zeros in $(I_n + A)y$ is less than k.

5.2.3 If A is an n-square indecomposable nonnegative matrix, then $(I_n + A)^{n-1}$ is positive.

5.2.4 If A is an indecomposable nonnegative matrix and x is a nonzero, nonnegative n-vector, then $Ax \neq 0_n$.

5.2.5 If an indecomposable nonnegative matrix has a positive characteristic root r and a corresponding nonnegative characteristic vector x, then $x > 0$.

5.2.6 Denote the (i,j) entry of A^k by $a_{ij}^{(k)}$. A nonnegative square matrix A is indecomposable, if and only if for each (i,j) there exists $k = k(i, j)$ such that $a_{ij}^{(k)} > 0$.

Note that k is a function of i,j and that for no k need all $a_{ij}^{(k)}$ be positive. For example, the permutation matrix P of 5.2.1 is indecomposable and yet, for any k, P^k contains $n^2 - n$ zeros.

5.2.7 If $A \geq 0$ is indecomposable, then for each (i,j) there exists an integer k, not greater than the degree of the minimal polynomial of A, such that $a_{ij}^{(k)} > 0$.

5.3 Examples

5.3.1 If a doubly stochastic matrix S is decomposable, then there exists a permutation matrix P such that PSP^T is a direct sum of doubly stochastic matrices.

For, if S is n-square, there exists, by the definition of decomposability, a permutation matrix P such that $PSP^T = \begin{pmatrix} B & 0 \\ C & D \end{pmatrix}$, where B is u-square and D is $(n - u)$-square. Let $\sigma(X)$ denote the sum of all the entries of the matrix X. Then $\sigma(B)$ is equal to the sum of entries in the first u rows of PSP^T and therefore $\sigma(B) = u$. Similarly, $\sigma(D) = n - u$. Hence $\sigma(PSP^T) = \sigma(B) + \sigma(C) + \sigma(D) = n + \sigma(C)$. But $\sigma(PSP^T) = n$ as $PSP^T \in \Omega_n$. Therefore $\sigma(C) = 0$ and, since $C \geq 0$, C must be a zero matrix and $PSP^T = B \dotplus D$. The matrices B and D are clearly doubly stochastic.

5.3.2 If $A \geq 0$ is indecomposable and $B > 0$, then $AB > 0$. For, if $(AB)_{ij} = 0$ for some i, j, then $\sum\limits_{k=1}^{n} a_{ik}b_{kj} = 0$ and, since all the b_{kj} are positive and the a_{ik} are nonnegative, we must have $a_{ik} = 0$, $(k = 1, \cdots, n)$. This contradicts the indecomposability of A.

5.4 Fully indecomposable matrices

If an n-square matrix A contains an $s \times (n - s)$ zero submatrix, for some integer s, then A is said to be *partly decomposable;* otherwise *fully indecomposable.* In other words, A is partly decomposable, if and only if there exist permutation matrices P and Q such that $PAQ = \begin{pmatrix} B & 0 \\ C & D \end{pmatrix}$ where B and D are square submatrices. Clearly every decomposable matrix is partly decomposable and every fully indecomposable matrix is indecomposable. For example, $\begin{pmatrix} 0 & 1 \\ 1 & 1 \end{pmatrix}$ is partly decomposable but not decomposable.

5.4.1 If $A \geq 0$ and $p(A) = 0$, then A is partly decomposable. For, by the Frobenius-König theorem (see 1.7.1), A must then contain an $s \times (n - s + 1)$ zero matrix.

5.4.2 A fully indecomposable n-square matrix can contain at most $n(n - 2)$ zeros. In fact, if P is the matrix in **5.2(1)**, then $I_n + P$ is fully indecomposable and therefore $I_n + P + B$ is also fully indecomposable for any nonnegative matrix B.

5.4.3 If A is fully indecomposable, then there exists a number k such that $A^p > 0$ for all $p \geq k$. The converse is not true; i.e., if an indecomposable matrix has the property that $A^p > 0$ for all $p \geq k$, it is not necessarily fully indecomposable. For example,

$$A = \begin{pmatrix} 0 & 1 \\ 1 & 1 \end{pmatrix}, \qquad A^2 > 0 \ .$$

5.4.4 If A is an n-square nonnegative matrix and AA^T contains an $s \times t$ zero submatrix such that $s + t = n - 1$, then A is partly decomposable. It follows that if A is fully indecomposable then so is AA^T.

5.4.5 If $A \geq 0$ is fully indecomposable and AA^T has t ($t > 0$) zeros in its ith row, then the number of zeros in the ith row of A is greater than t.

5.4.6 If $A \geq 0$ is a fully indecomposable n-square matrix, then no row of AA^T can have more than $n - 3$ zeros.

5.5 Perron-Frobenius theorem

We now state the fundamental theorem on nonnegative matrices. It is the generalized and amplified version, due to Frobenius, of Perron's theorem.

5.5.1 Let A be an n-square nonnegative indecomposable matrix. Then:

(i) A has a real positive characteristic root r (the *maximal root of A*) which is a simple root of the characteristic equation of A. If λ_i is any characteristic root of A, then $|\lambda_i| \leq r$.

(ii) There exists a positive characteristic vector corresponding to r.

(iii) If A has h characteristic roots of modulus r: $\lambda_0 = r, \lambda_1, \cdots, \lambda_{h-1}$, then these are the h distinct roots of $\lambda^h - r^h = 0$; h is called the *index of imprimitivity* of A.

(iv) If $\lambda_0, \lambda_1, \cdots, \lambda_{n-1}$ are all the characteristic roots of A and $\theta = e^{i2\pi/h}$, then $\lambda_0\theta, \cdots, \lambda_{n-1}\theta$ are $\lambda_0, \lambda_1, \cdots, \lambda_{n-1}$ in some order.

(v) If $h > 1$, then there exists a permutation matrix P such that

$$PAP^T = \begin{pmatrix} 0 & A_{12} & 0 & \cdots & 0 & 0 \\ 0 & 0 & A_{23} & \cdots & 0 & 0 \\ \multicolumn{6}{c}{\cdots\cdots\cdots\cdots\cdots\cdots\cdots\cdots\cdots\cdots} \\ \multicolumn{6}{c}{\cdots\cdots\cdots\cdots\cdots\cdots\cdots\cdots\cdots\cdots} \\ 0 & 0 & 0 & \cdots & 0 & A_{h-1,h} \\ A_{h,1} & 0 & 0 & \cdots & 0 & 0 \end{pmatrix}$$

where the zero blocks down the main diagonal are square.

Any proof of the Perron-Frobenius theorem is too long and too involved to be given here. We list, however, some important related results.

5.5.2 Let $\mathbf{P}^n$ denote the set of n-tuples with nonnegative entries. For a nonzero $x \in \mathbf{P}^n$ and a fixed nonnegative indecomposable n-square matrix A define:

$$r(x) = \min_i \frac{(Ax)_i}{x_i}, \qquad (x_i \neq 0), \tag{1}$$

$$s(x) = \max_i \frac{(Ax)_i}{x_i}, \qquad (x_i \neq 0). \tag{2}$$

In other words, $r(x)$ is the largest real number α for which $Ax - \alpha x \geq 0$ and $s(x)$ is the least real number β for which $Ax - \beta x \leq 0$. Then for any nonzero $x \in \mathbf{P}^n$

$$r(x) \leq r \leq s(x). \tag{3}$$

Also

$$r = \max_{x \in \mathbf{P}^n} r(x) = \min_{x \in \mathbf{P}^n} s(x). \tag{4}$$

5.5.3 If $A \geq 0$ is indecomposable, r is the maximal root of A and $B(x) = \text{adj}\,(xI_n - A)$, $(x \in \mathbf{R})$, then $B(x) > 0$ for $x \geq r$. Moreover $(xI_n - A)^{-1} > 0$ for $x > r$.

5.5.4 If $A \in M_n(\mathbf{R})$ is the matrix defined above and $C(\lambda)$ is the reduced adjoint matrix of $\lambda I_n - A$, i.e., $C(\lambda) = B(\lambda)/f_{n-1}$ where f_{n-1} is the gcd of all $(n-1)$-square subdeterminants of $\lambda I_n - A$ (see **I.3.15**), then $C(r) > 0$. Here λ is of course an indeterminate over $\mathbf{R}$ and $C(r)$ is the matrix obtained from $C(\lambda)$ by replacing λ by the maximal root r of A.

5.5.5 If $A \geq 0$ is indecomposable, then A cannot have two linearly independent nonnegative characteristic vectors.

5.5.6 If $A \geq 0$ is an indecomposable matrix in $M_n(\mathbf{R})$, then the maximal characteristic root of $A(\omega|\omega)$, $(\omega \in Q_{k,n}, 1 \leq k < n)$, is less than the maximal characteristic root of A.

5.6 Example

Let $u = (u_1, \cdots, u_n)$ and define $\sigma(u)$ to be the sum $\sum\limits_{j=1}^{n} u_j$. Let $A \geq 0$ be an indecomposable matrix with maximal root r and (unique) characteristic vector x corresponding to r chosen so that $\sigma(x) = 1$. If $R \geq 0$ is any nonzero matrix commuting with A, then $Rx/\sigma(Rx) = x$.

For, $Ax = rx$ and therefore $RAx = rRx$; i.e., $A(Rx) = r(Rx)$. Since $Rx \neq 0$, it must be a characteristic vector of A corresponding to r and therefore a nonzero multiple of x (see **5.5.5**). Hence $Rx/\sigma(Rx) = x$.

5.7 Nonnegative matrices

For nonnegative matrices which are not necessarily indecomposable the following versions of the results in **5.5** hold.

5.7.1   If $A \geq 0$, then A has a maximal nonnegative characteristic root r such that $r \geq |\lambda_j|$ for any characteristic root λ_j of A.

5.7.2   If $A \geq 0$, then there exists a nonnegative characteristic vector corresponding to the maximal characteristic root of A.

5.7.3   If $A \geq 0$ and $B(x) = \text{adj } (xI_n - A)$, then $B(x) = (b_{ij}(x)) \geq 0$ for $x \geq r$; moreover, $db_{ij}(x)/dx \geq 0$ for each i,j and $x \geq r$, where r denotes the maximal root of A.

5.7.4   If $C(\lambda)$ is the reduced adjoint matrix of $\lambda I_n - A$ (see **5.5.4**), then $C(\lambda) \geq 0$ for $\lambda \geq r$.

We list below some further related results on nonnegative matrices.

5.7.5   If $A \geq 0$ has maximal root r and γ is any characteristic root of a matrix $G \in M_n(\mathbf{C})$ for which $|g_{ij}| \leq a_{ij}$, then $|\gamma| \leq r$. If, in addition, A is indecomposable and $|\gamma| = r$, i.e., $\gamma = re^{i\varphi}$, then $G = e^{i\varphi}DAD^{-1}$ where $|d_{ij}| = \delta_{ij}$, $(i, j = 1, \cdots, n)$.

5.7.6   If $A \geq 0$ has maximal root r and $k \leq n$, then the maximal root of $A(\omega|\omega)$, $(\omega \in Q_{k,n}, 1 \leq k < n)$, does not exceed r.

5.7.7   Let A and $B(x)$ be the matrices defined in **5.7.3**. Then A is decomposable if and only if $B(r)$ has a zero entry on its main diagonal.

5.7.8   Let A be the matrix in **5.7.3**. Then every principal subdeterminant of $xI_n - A$ is positive if and only if $x > r$.

5.7.9   If $A \geq 0$ has a simple maximal root r and both A and A^T have positive characteristic vectors corresponding to r, then A is indecomposable.

5.8 Examples

5.8.1   Let A and B be nonnegative matrices and suppose $A * B$ is their Hadamard product (see **4.5**). Let $\lambda_M(A)$, $\lambda_M(B)$, $\lambda_M(A * B)$ denote the maximal roots of A, B, and $A * B$ respectively. Then

$$\lambda_M(A * B) \leq \lambda_M(A)\lambda_M(B).$$

For, by **4.5.1**, $A * B$ is a principal submatrix of $A \otimes B$ and the maximal root of $A \otimes B$ is $\lambda_M(A)\lambda_M(B)$ (see **I.2.15.11**). It follows therefore from **5.7.6** that $\lambda_M(A * B) \leq \lambda_M(A)\lambda_M(B)$.

5.8.2   Let $A = (a_{ij}) \in M_n(\mathbf{R})$ be a nonnegative matrix with a maximal root r not exceeding 1. Then a necessary and sufficient condition that there exist a permutation matrix P for which PAP^T is triangular is that

$$\prod_{i=1}^{n} (1 - a_{ii}) = d(I_n - A). \tag{1}$$

For, the condition is clearly necessary. To prove the sufficiency consider a nonnegative matrix $C = (c_{ij}) \in M_m(\mathbf{R})$, with a maximal root s not exceeding 1, and consider the function

$$f(t) = \prod_{i=1}^{m} (t - c_{ii}) - d(tI_m - C). \tag{2}$$

We use induction on m to show that $f(t) \geq 0$ for $t \geq s$. Differentiate 5.8(2),

$$f'(t) = \sum_{i=1}^{m} \left\{ \prod_{j \neq 1} (t - c_{jj}) - d(tI_{m-1} - C(i|i)) \right\}. \tag{3}$$

Since $t \geq s$ and s is not smaller than the maximal root of $C(i|i)$ (see **5.7.6**), each summand in **5.8(3)** is nonnegative by the induction hypothesis. Thus $f'(t) \geq 0$ for $t \geq s$. Now,

$$f(s) = \prod_{i=1}^{m} (s - c_{ii}) \geq 0,$$

since $s \geq c_{ii}$, $(i = 1, \cdots, m)$, by **5.7.6**. Hence $f(t) \geq 0$ for $t \geq s$. Moreover, if C is indecomposable then $s > c_{ii}$, $(i = 1, \cdots, m)$, (see **5.5.6**) and $f(s) > 0$ and thus $f(t) > 0$ for $t \geq s$. It follows that if C is indecomposable, then $f(1) \geq f(s) > 0$; i.e.,

$$\prod_{i=1}^{m} (1 - c_{ii}) > d(I_m - C).$$

Now, choose a permutation matrix P such that $PAP^T = B = \sum_{i=1}^{m} \cdot B_i$, where each B_i is 1-square or possibly n_i-square indecomposable. We show that the second alternative is impossible and therefore that PAP^T is triangular. Since the b_{ii} are just a permutation of the a_{ii}, **5.8(1)** implies that

$$\prod_{i=1}^{n} (1 - a_{ii}) = \prod_{i=1}^{n} (1 - b_{ii}) = d(I_n - A) = d(I_n - B) = \prod_{k=1}^{m} d(I_{n_k} - B_k).$$

Suppose that there exists a q such that $B_q = C$ is indecomposable. Then, by the above argument,

$$\prod_{i=1}^{n_k} (1 - c_{ii}) > d(I_{n_q} - B_q)$$

and therefore

$$\prod_{k=1}^{m} d(I_{n_k} - B_k) < \prod_{i=1}^{n} (1 - a_{ii}),$$

a contradiction.

5.9 Primitive matrices

Let A be an indecomposable nonnegative matrix with maximal root r. Let $h(A)$ be the number of characteristic roots of A each of which has the property that its modulus is equal to r. Then A is said to be *primitive*, if $h(A) = 1$. We recall that the number $h(A)$ is called the index of imprimitivity of A (see **5.5.1**).

5.9.1   If $A \geq 0$ is an indecomposable matrix with characteristic polynomial

$$\lambda^n + a_1\lambda^{n_1} + \cdots + a_t\lambda^{n_t}, \qquad (a_i \neq 0),$$

then

$$h(A) = \gcd(n - n_1, n_1 - n_2, \cdots, n_{t-1} - n_t). \tag{1}$$

5.9.2 $A \geq 0$ is primitive if and only if $A^p > 0$ for some positive integer p. Moreover, if $A^p > 0$, then $A^q > 0$ for $q \geq p$.

5.9.3   If $A \geq 0$ is primitive, then A^q is primitive for any positive integer q.

5.9.4   If $A \geq 0$ is an indecomposable matrix with index of imprimitivity h, then there exists a permutation matrix P such that

$$PA^hP^T = \sum_{i=1}^{h} A_i$$

where the A_i are primitive matrices with the same maximal root.

5.9.5   A fully indecomposable matrix is primitive (see **5.4.5**). The converse is not true.

5.9.6   If $A \geq 0$ is primitive, then there exists a positive integer p, $(1 \leq p \leq (n-1)^2 + 1)$, such that $A^p > 0$. If $a_{ii} > 0$, $(i = 1, \cdots, n)$, then $A^{n-1} > 0$.

5.9.7   If $A \geq 0$ is primitive with maximal root r (see **5.5.1**), then $\lim_{m \to \infty} \left(\dfrac{A}{r}\right)^m = B$ where $\rho(B) = 1$ and $B^{(i)}$ is a positive multiple of the positive characteristic vector of A corresponding to r. It follows from this that

$$\lim_{m \to \infty} (a_{ij}^{(m)})^{1/m} = r, \qquad (i, j = 1, \cdots, n). \tag{2}$$

Recall (see **5.2.6**) that $a_{ij}^{(m)}$ is the (i,j) entry of A^m. Moreover, if $A \geq 0$ is assumed only to be indecomposable, then **5.9(2)** is equivalent to A being primitive.

5.10 Example

5.10.1   Let P be the full-cycle permutation matrix in **5.2(1)** and $E_{i,i}$ the matrix with 1 in the (i,i) position and zero elsewhere. Then $P + E_{i,i}$ is primitive. For, $P + E_{i,i}$ is an indecomposable matrix with a positive trace. Therefore in **5.9(1)** the number $n - n_1$ is equal to 1 and thus $h = 1$.

More generally, if A is any indecomposable matrix and B any nonnegative matrix with a positive trace, then $A + B$ is primitive.

5.10.2   The least integer p for which a primitive matrix $A \geq 0$ satisfies $A^p > 0$ is not in general less than $(n - 1)^2 + 1$ (see **5.9.6**). For, consider the primitive matrix $A = P + E_{n,2}$ where P is the full-cycle permutation matrix in **5.2.1** and $E_{n,2}$ is the matrix with 1 in position $(n,2)$ and 0 elsewhere. Then A^k contains zero entries for $k < (n - 1)^2 + 1$.

5.11   Doubly stochastic matrices

A matrix $A \geq 0$ is doubly stochastic, if all its row sums and all its column sums are equal to 1 (see **1.4.2**). Clearly a nonnegative matrix $A \in M_n(\mathbf{R})$ is doubly stochastic, if and only if $J_n A = A J_n = J_n$ where $J_n \in M_n(\mathbf{R})$ is the matrix all of whose entries are equal to $1/n$. For example, permutation matrices as well as matrices of the form $k^{-1}V$ where V is a (v,k,λ)-matrix (see **I.2.4.10**) are doubly stochastic.

5.11.1   The set of all n-square doubly stochastic matrices, Ω_n, forms a convex polyhedron with the permutation matrices as vertices (see **1.7**).

5.11.2   If $S \in \Omega_n$, then for each integer k, $(1 \leq k \leq n)$, there exists a diagonal d of S containing at least $n - k + 1$ elements not smaller than μ, where

$$\mu = \begin{cases} \dfrac{4k}{(n + k)^2} & \text{if } n + k \text{ is even,} \\[3mm] \dfrac{4k}{(n + k)^2 - 1} & \text{if } n + k \text{ is odd.} \end{cases}$$

5.11.3   If $S \in \Omega_n$ and more than $(n - 1)(n - 1)!$ terms in the expansion of $p(S)$ have a common nonzero value, then $A = J_n$.

5.11.4   Every $S \in \Omega_n$ can be expressed as a convex combination of fewer than $(n - 1)^2 + 2$ permutation matrices.

5.11.5   If $S \in \Omega_n$, then

$$((n - 1)^2 + 1)^{1-n} \leq p(S) \leq 1.$$

The upper bound is attained, if and only if S is a permutation matrix.

5.11.6   (van der Waerden conjecture)   It is conjectured that if $S \in \Omega_n$, then

$$p(S) \geq \frac{n!}{n^n}$$

with equality, if and only if $S = J_n$. This conjecture is still unresolved. Some partial answers are given below.

5.11.7 If $A \in \Omega_n$ and $p(A) = \min_{S \in \Omega_n} p(S)$, then

(i) $p(A(i|j)) = p(A)$, if $a_{ij} \neq 0$,

(ii) $p(A(i|j)) = p(A) + \beta$, if $a_{ij} = 0$ where $\beta \geq 0$ is independent of i,j.

5.11.8 If $A \in \Omega_n$ satisfies $p(A) = \min_{S \in \Omega_n} p(S)$, then A is fully indecomposable.

5.11.9 If $A \in \Omega_n$ satisfies $p(A) = \min_{S \in \Omega_n} p(S)$ and $A > 0$, then $A = J_n$.

5.11.10 If $A \in \Omega_n$, $(A \neq J_n)$, and the elements of $A - J_n$ are sufficiently small in absolute value, then $p(A) > p(J_n)$.

5.11.11 If $A \in \Omega_n$ is indecomposable, then its index of imprimitivity, h, divides n. Moreover, there exist permutation matrices P and Q such that

$$PAQ = \sum_{j=1}^{h} {}^{\cdot} A_j$$

where $A_j \in \Omega_{n/h}$, $(j = 1, \cdots, h)$.

5.11.12 By the Birkhoff theorem (see **1.7**), every $A \in \Omega_n$ can be represented (in general, not uniquely) as a convex combination of permutation matrices. Let $\beta(A)$ be the least number of permutation matrices necessary to represent $A \in \Omega_n$ as such a combination. Then, if A is indecomposable with index of imprimitivity h,

$$\beta(A) \leq h\left(\frac{n}{h} - 1\right)^2 + 1.$$

5.11.13 If $A \in \Omega_n$ and P and Q are permutation matrices such that

$$PAQ = \sum_{j=1}^{m} {}^{\cdot} A_j, \tag{1}$$

where each A_j is n_j-square primitive or is 1-square, then

$$\beta(A) \leq \sum_{j=1}^{m} (n_j - 1)^2 + 1.$$

It is clear that any $A \in \Omega_n$ can be written in the form **5.11(1)**. For any given $A \in \Omega_n$ the numbers n_j are uniquely determined (except for order).

5.11.14 If A is a positive semidefinite symmetric nonnegative matrix without a zero row, then there exists a real diagonal matrix D such that DAD is doubly stochastic. Moreover, if G is a diagonal matrix such that GAG is doubly stochastic, then $DAD = GAG$.

5.11.15 If $S = (s_{ij}) \in \Omega_n$ and $A = (a_{ij}) \in M_n(\mathbf{R})$ is a symmetric positive semidefinite matrix such that $a_{ij} \geq s_{ij}$, $(i,j = 1, \cdots, n)$, then $p(A) \geq n!/n^n$.

The equality holds only if $A = QJ_nQ$ where Q is a diagonal matrix whose permanent is 1.

5.12 Examples

5.12.1 If $S = (s_{ij}) \in \Omega_n$, then there exists a permutation $\sigma \in S_n$ such that

$$\prod_{i=1}^{n} s_{i,\sigma(i)} \geq \frac{1}{n^n} \tag{1}$$

with equality, if and only if $S = J_n$.

For, let F be the convex function on Ω_n defined in **2.4.2**. Then if P is a full cycle permutation matrix,

$$\begin{aligned} F(S) &= \frac{1}{n} \sum_{\alpha=0}^{n-1} F(P^\alpha S) \\ &\geq F\left(\sum_{\alpha=0}^{n-1} \frac{P^\alpha S}{n} \right) \\ &= F(J_n S) \\ &= F(J_n) \\ &= n \log\left(\frac{1}{n} \right) \end{aligned} \tag{2}$$

with equality, if and only if $P^\alpha S = S$ for all α; i.e., if and only if $S = J_n$. The first equality in **5.12(2)** follows from the fact that $F(QS) = F(S)$ for any permutation matrix Q.

By **1.7**, there exist permutations $\sigma \in S_n$ such that $\prod_{i=1}^{n} s_{i,\sigma(i)} > 0$. Hence $\max_{\sigma} \prod_{i=1}^{n} s_{i,\sigma(i)} > 0$ and we can consider $\max_{\sigma \in S_n} \sum_{i=1}^{n} \log s_{i,\sigma(i)}$. Let $P = (p_{ij})$ be a permutation matrix in Ω_n. Clearly

$$\max_{\sigma \in S_n} \sum_{i=1}^{n} \log s_{i,\sigma(i)} = \max_{P} \sum_{i,j} p_{ij} \log s_{ij}.$$

Now, let $T = (t_{ij})$ be any matrix in Ω_n. Then $\sum_{i,j} t_{ij} \log s_{ij}$ is linear in T and therefore takes its maximum on a permutation matrix, a vertex of the convex polyhedron Ω_n (see **1.12**). Thus

$$\begin{aligned} \max_{P \in \Omega_n} \sum p_{ij} \log s_{ij} &= \max_{T \in \Omega_n} \sum_{i,j} t_{ij} \log s_{ij} \\ &\geq \sum_{i,j} f(s_{ij}) \\ &= F(S) \\ &\geq n \log\left(\frac{1}{n} \right). \end{aligned}$$

It follows that $\max\limits_{\sigma \in S_n} \prod\limits_{i=1}^{n} s_{i,\sigma(i)} \geq 1/n^n$ with equality, if and only if $S = J_n$.

For, if $\max\limits_{\sigma \in S_n} \prod\limits_{i=1}^{n} s_{i,\sigma(i)} = 1/n^n$, then $F(S) = n \log (1/n)$ and thus $S = J_n$.

Conversely, if $S = J_n$, then clearly $\prod\limits_{i=1}^{n} s_{i,\sigma(i)} = 1/n^n$ for all $\sigma \in S_n$.

Observe that the above result is implied by the van der Waerden conjecture (see **5.11.6**).

5.12.2 If $A \in \Omega_n$ is positive semidefinite symmetric, then there exists a positive semidefinite real symmetric matrix B such that $BJ = JB = J$ and $B^2 = A$.

Observe first that A has 1 as a characteristic root with $(1/\sqrt{n}, \cdots, 1/\sqrt{n})$ as corresponding characteristic vector. Hence, by **I.4.10.5** there exists a real orthogonal matrix U such that $A = U \operatorname{diag}(1, \lambda_2, \cdots, \lambda_n)U^T$. Clearly all elements in the first column of U are $1/\sqrt{n}$. Let

$$B = U \operatorname{diag}(1, \sqrt{\lambda_2}, \cdots, \sqrt{\lambda_n})U^T.$$

Then B is positive semidefinite real symmetric and $B^2 = A$. It remains to prove that $BJ = JB = J$. Since $J^{(k)} = \sqrt{n}U^{(1)}$ and $U^T U^{(1)} = e_1$, we have $U^T J^{(k)} = \sqrt{n}U^T U^{(1)} = \sqrt{n}e_1$ and therefore

$$\begin{aligned}
BJ^{(k)} &= U \operatorname{diag}(1, \sqrt{\lambda_2}, \cdots, \sqrt{\lambda_n})U^T J^{(k)} \\
&= \sqrt{n}U \operatorname{diag}(1, \sqrt{\lambda_2}, \cdots, \sqrt{\lambda_n})e_1 \\
&= \sqrt{n}Ue_1 \\
&= \sum_{j=1}^{n} e_j
\end{aligned}$$

and thus $BJ = J$. Also $JB = (BJ)^T = J^T = J$.

5.12.3 If $A \in \Omega_n$ is positive semidefinite, then

$$p(A) \geq \frac{n!}{n^n}$$

with equality, if and only if $A = J_n$.

By **5.12.2**, there exists a real symmetric matrix B such that $B^2 = A$ and $JB = J$. The inequality **4.4(3)** gives

$$(p(JB))^2 \leq p(JJ^*)p(B^*B).$$

Now $JB = J$, $JJ^* = nJ$, and $B^*B = B^2 = A$. Therefore

$$(p(J))^2 \leq n^n p(J)p(A)$$

and

$$p(A) \geq \frac{p(J)}{n^n} = \frac{n!}{n^n}.$$

We omit the proof of the case of equality.

5.12.4   If $A \in \Omega_n$ is positive semidefinite symmetric with $a_{ii} \leq 1/(n-1)$, $(i = 1, \cdots, n)$, then there exists a positive semidefinite symmetric $B \in \Omega_n$ such that $B^2 = A$.

We show that the condition $a_{ii} \leq 1/(n-1)$, $(i = 1, \cdots, n)$, implies that the matrix B in **5.12.2** satisfies $B \geq 0$. For, suppose that $b_{i_0 j_0} < 0$ for some i_0, j_0. Then $\sum_{j \neq j_0} b_{i_0 j} = \mu > 1$ since $\sum_{j=1}^{n} b_{i_0 j} = 1$. Then

$$\mu^2 = \left(\sum_{j \neq j_0} b_{i_0 j} \right)^2 \leq \left(\sum_{j \neq j_0} b_{i_0 j}^2 \right) (n-1),$$

by **3.1(4)**. Thus

$$a_{i_0 i_0} = \sum_{j=1}^{n} b_{i_0 j}^2 \geq \sum_{j \neq j_0} b_{i_0 j}^2 \geq \frac{\mu^2}{n-1} > \frac{1}{n-1},$$

a contradiction.

5.13   Stochastic matrices

A matrix $A \in M_n(\mathbf{R})$, $A \geq 0$, is *row stochastic*, if $AJ = J$ where again J is the matrix in $M_n(\mathbf{R})$ all of whose entries are 1. In other words, every row sum of A is 1.

5.13.1   The set of row stochastic matrices is a polyhedron whose vertices are the n^n matrices with exactly one entry 1 in each row.

5.13.2   The matrix $A \geq 0$ is row stochastic, if and only if the vector $(1, \cdots, 1)$ is a characteristic vector of A corresponding to the characteristic root 1.

5.13.3   If A is row stochastic, then every characteristic root r of A satisfies $|r| \leq 1$.

5.13.4   If A is row stochastic, then $\lambda - 1$ appears only linearly in the list of elementary divisors of $\lambda I_n - A$ [i.e., $(\lambda - 1)^k$, $(k > 1)$, is never an elementary divisor of $\lambda I_n - A$].

5.13.5   If A is nonnegative and indecomposable with maximal root r, then there exists a diagonal matrix

$$D = \operatorname{diag}(d_1, \cdots, d_n), \quad (d_i > 0; i = 1, \cdots, n),$$

such that $r^{-1}D^{-1}AD$ is row stochastic.

References

§1.4.4. This example is due to A. J. Hoffman ([B12]).

§1.7. [B1].

§1.7.1. [A7], p. 240; [B7].

§3.1.1. [A6], p. 26.

§3.1.3. This is Beckenbach's inequality ([A1], p. 27).

§§3.2.1, 3.2.2. [A6], p. 52.

§§3.2.3, 3.2.4. [B2].

§3.2.5. [B17].

§3.2.6. [B2].

§3.2.7. [B31].

§3.3. [A6], p. 26; [A1], p. 19.

§3.4. [A6], p. 31.

§3.5.1. [A6], p. 21; [A1], p. 26.

§3.5.2. [B9].

§3.5.3. [A1], p. 3.

§3.5.4. [A6], p. 47.

§3.6. The solution follows the method of proof of Lemma 2 in [A6], p. 47.

§§4.1.1–4.1.4. [B10].

§4.1.5. [B5].

§4.1.6. For the lower bound see [B5]; for the upper bound see [B18].

§4.1.7. [A1], p. 63; [B14].

§4.1.8. *Ibid.*, p. 70.

§4.1.9. [B30]; [B28].

§4.1.10. For inequality 4.1(10) see [B2]; for case $k = 1$ see also [B17].

§4.2. [B30]; [B4], I; [B12].

§4.3.1. [B9].

§4.3.2. [B15].

§4.4.2. This inequality is due to E. Fischer; [A8], p. 420.

§§4.4.3, 4.4.4. [B23].

§4.4.5. For inequality 4.4(5) see [B18].

§4.4.6. This is Szász's inequality; [A1], p. 64.

§4.4.7. [A8], p. 75.

§4.4.8. This is Bergstrom's inequality; [A1], p. 67.

§4.4.9. This is a generalization of Heinz's inequality; [B13].

§4.4.10. [B8].

§§4.4.11–4.4.13. [B11].

§4.4.14. [B33].

§4.4.15. This is an unpublished result of Ky Fan.

§4.4.16. These are Aronszajn's inequalities; [A5], p. 77.

§§4.5.1–4.5.3. [B16].

§4.5.4. This inequality is due to Schur; [A8], p. 422.

§§5.2.2–5.2.7. [A4], p. 50.

§§5.4.2–5.4.6. [B20].

§5.5.1. [A4], p. 53.

§5.5.2. [B3]; [A1], p. 81.

§§5.5.3–5.5.6. [A4], ch. XIII.

§5.6. [B21].

§§5.7.1–5.7.9. [A4], ch. XIII.

§5.8.1. [B16].

§5.8.2. [B21].

§§5.9.1–5.9.4. [A4], p. 80.

§5.9.6. *Ibid.;* [B32].

§5.9.7. [A4], p. 81.

§5.10.2. [B32].

§5.11.1. [B1].

§5.11.2. [B19]; [B24].

§5.11.3. [B19].

§5.11.4. [B24].

§5.11.6. [B29]; [B22].

§§5.11.7–5.11.10. [B22].

§§5.11.11–5.11.13. [B21].

§§5.11.14, 5.11.15. These unpublished results are due to M. Marcus and M. Newman; see also [B25].

§5.12.1. [B19].

§5.12.2. *Ibid.;* [B26].

§5.12.3. [B23]; [B26].

§5.12.4. [B19].

§§5.13.2–5.13.5. [A4], p. 82.

Bibliography

Part A. Books

1. Beckenbach, E. F., and Bellman, R., *Inequalities*, Springer, Berlin (1961).

2. Bonnesen, T., and Fenchel, W., *Theorie der konvexen Körper*, Chelsea, New York (1948).

3. Eggleston, H. G., *Convexity*, Cambridge University, London (1958).

4. Gantmacher, F. R., *The theory of matrices*, vol. II, Chelsea, New York (1959).

5. Hamburger, H. L., and Grimshaw, M. E., *Linear transformations in n-dimensional vector space*, Cambridge University, London (1951).

6. Hardy, G. H., Littlewood, J. E., and Pólya, G., *Inequalities*, 2nd ed., Cambridge University, London (1952).

7. König, D., *Theorie der endlichen und unendlichen Graphen*, Chelsea, New York (1950).

8. Mirsky, L., *An introduction to linear algebra*, Oxford University, Oxford (1955).

Part B.   Papers

1. Birkhoff, G., *Tres observaciones sobre el algebra lineal*, Univ. Nac. Tucumán Rev. Ser. A, *5* (1946), 147–150.

2. Bullen, P., and Marcus, M., *Symmetric means and matrix inequalities*, Proc. Amer. Math. Soc., *12* (1961), 285–290.

3. Collatz, L., *Einschliessungssatz für die charakteristischen Zahlen von Matrizen*, Math. Z., *48* (1942), 221–226.

4. Fan, Ky, *On a theorem of Weyl concerning eigenvalues of linear transformations*, Proc. Nat. Acad. Sci., I, *35* (1949), 652–655; II, *36* (1950), 31–35.

5. Fan, Ky, *A minimum property of the eigenvalues of a hermitian transformation*, Amer. Math. Monthly, *60* (1953), 48–50.

6. Frobenius, G., *Über Matrizen aus positiven Elementen*, S.-B. Kgl. Preuss. Akad. Wiss., I (1908), 471–476; II (1909), 514–518.

7. Frobenius, G., *Über Matrizen aus nicht negativen Elementen*, S.-B. Kgl. Preuss. Akad. Wiss. (1912), 456–477.

8. Hua, L. K., *Inequalities involving determinants* (in Chinese), Acta Math. Sinica, *5* (1955), 463; Math. Rev., *17* (1956), 703.

9. Kantorovich, L. V., *Functional analysis and applied mathematics*, Uspehi Mat. Nauk, *3* (1948), 89–185 [English trans.: Nat. Bur. Standards, Report No. 1509 (1952)].

10. Marcus, M., *Convex functions of quadratic forms*, Duke Math. J., *24* (1957), 321–325.

11. Marcus, M., *On subdeterminants of doubly stochastic matrices*, Illinois J. Math., *1* (1957), 583–590.

12. Marcus, M., *Some properties and applications of doubly stochastic matrices*, Amer. Math. Monthly, *67* (1960), 215–221.

13. Marcus, M., *Another extension of Heinz's inequality*, J. Res. Nat. Bur. Standards, *65B* (1961), 129–130.

14. Marcus, M., *The permanent analogue of the Hadamard determinant theorem*, Bull. Amer. Math. Soc., *4* (1963), 494–496.

15. Marcus, M., and Khan, N. A., *Some generalizations of Kantorovich's inequality*, Portugal. Math., *20* (1951), 33–38.

16. Marcus, M., and Khan, N. A., *A note on the Hadamard product*, Canad. Math. Bull., *2* (1959), 81–83.

17. Marcus, M., and Lopes, L., *Inequalities for symmetric functions and hermitian matrices*, Canad. J. Math., *9* (1957), 305–312.

18. Marcus, M., and McGregor, J. L., *Extremal properties of hermitian matrices*, Canad. J. Math., *8* (1956), 524–531.

19. Marcus, M., and Minc, H., *Some results on doubly stochastic matrices*, Proc. Amer. Math. Soc., *13* (1962), 571–579.

20. Marcus, M., and Minc, H., *Disjoint pairs of sets and incidence matrices*, Illinois J. Math., *7* (1963), 137–147.

21. Marcus, M., Minc, H., and Moyls, B., *Some results on nonnegative matrices*, J. Res. Nat. Bur. Standards, *65B* (1961), 205–209.

22. Marcus, M., and Newman, M., *On the minimum of the permanent of a doubly stochastic matrix*, Duke Math. J., *26* (1959), 61–72.

23. Marcus, M., and Newman, M., *Inequalities for the permanent function*, Ann. of Math., *75* (1962), 47–62.

24. Marcus, M., and Ree, R., *Diagonals of doubly stochastic matrices*, Quart. J. Math. Oxford Ser. (2), *10* (1959), 295–302.

25. Maxfield, J. E., and Minc, H., *On the doubly stochastic matrix diagonally congruent to a given matrix* (submitted for publication).

26. Minc, H., *A note on an inequality of M. Marcus and M. Newman*, Proc. Amer. Math. Soc., *14* (1963), 890–892.

27. Mirsky, L., *Results and problems in the theory of doubly stochastic matrices*, Z. Wahrscheinlichkeitstheorie *1* (1963), 319–334.

28. Pólya, G., *Remark on Weyl's note: "Inequalities between the two kinds of eigenvalues of a linear transformation,"* Proc. Nat. Acad. Sci., *36* (1950), 49–51.

29. van der Waerden, B. L., *Aufgabe 45*, Jber. Deutsch. Math. Verein., *35* (1926), 117.

30. Weyl, H., *Inequalities between the two kinds of eigenvalues of a linear transformation*, Proc. Nat. Acad. Sci., *35* (1949), 408–411.

31. Whiteley, J. N., *Some inequalities concerning symmetric functions*, Mathematika, *5* (1958), 49–57.

32. Wielandt, H., *Unzerlegbare, nicht negative Matrizen*, Math. Z., *52* (1950), 642–648.

33. Wielandt, H., *An extremum property of sums of eigenvalues*, Proc. Amer. Math. Soc., *6* (1955), 106–110.

Localization of Characteristic Roots

III

1 Bounds for Characteristic Roots

1.1 Introduction

Unless otherwise stated, all matrices are assumed to be square and over the complex field, i.e., in $M_n(\mathbf{C})$.

There is quite a lot of information about the characteristic roots of some special types of matrices. Thus diagonal and triangular matrices exhibit their characteristic roots on their main diagonal (see **I.2.15.2**). All characteristic roots of nilpotent matrices are equal to 0 while all those of idempotent matrices (i.e., $A^2 = A$) are equal to 0 or 1. At least one of the characteristic roots of any row stochastic matrix is equal to 1 and all of them lie on or inside the unit circle. All the characteristic roots of a unitary matrix lie on the unit circle, those of a hermitian matrix lie on the real axis, while those of a skew-hermitian matrix lie on the imaginary axis (see **I.4.7.19, I.4.7.18, I.4.7.20**).

As far as the characteristic roots of a general matrix are concerned nothing specific can be said about their location: they can obviously lie anywhere in the complex plane. In this chapter we shall state and prove theorems on bounds for characteristic roots of a general matrix in terms of simple functions of its entries or of entries of a related matrix as well as theorems on bounds for characteristic roots of other classes of matrices.

In the first section we shall discuss some results obtained in the first decade of this century. The first results that specifically give bounds for characteristic roots of a general (real) matrix are due to Ivar Bendixson and are dated 1900. It is true that a simple corollary to the so-called Hadamard theorem on nonsingularity of square matrices [which incidentally is due to L. Lévy (1881) and in its general form to Desplanques (1887)] gives the well-known Geršgorin result. However, that simple corollary was not enunciated until 1931 (in this country not until 1946) and the honor of being the first in the field of localization of characteristic roots must be Bendixson's.

We shall use the standard notation of matrix theory and in **1.2, 1.3,** and **1.4** the following special notation: If $A = (a_{ij}) \in M_n(\mathbf{C})$, let $B = (b_{ij}) = (A + A^*)/2$ and $C = (c_{ij}) = (A - A^*)/2i$, both of which are obviously hermitian. Let $\lambda_1, \cdots, \lambda_n, (|\lambda_1| \geq \cdots \geq |\lambda_n|), \mu_1 \geq \cdots \geq \mu_n, \nu_1 \geq \cdots \geq \nu_n$, be the characteristic roots of A, B, C respectively and let $g = \max_{i,j} |a_{ij}|$, $g' = \max_{i,j} |b_{ij}|$, $g'' = \max_{i,j} |c_{ij}|$.

1.2 Bendixson's theorems

1.2.1 If $A \in M_n(\mathbf{R})$, then

$$|Im(\lambda_t)| \leq g'' \sqrt{\frac{n(n-1)}{2}}.$$

1.2.2 If $A \in M_n(\mathbf{R})$, then

$$\mu_n \leq Re(\lambda_t) \leq \mu_1.$$

We shall not prove Bendixson's theorems here as we shall presently state and prove their generalizations to complex matrices by A. Hirsch. Hirsch's paper in the *Acta Mathematica* follows the French version of Bendixson's paper in that journal and is in the form of a letter to Bendixson. Its style is somewhat amusing because of rather exaggerated eulogistic references to Bendixson's results and methods.

1.3 Hirsch's theorems

1.3.1 If $A \in M_n(\mathbf{C})$, then

$$|\lambda_t| \leq ng,$$
$$|Re(\lambda_t)| \leq ng',$$
$$|Im(\lambda_t)| \leq ng''.$$

If $A + A^t$ is real, then this last inequality can be replaced by

$$|Im(\lambda_t)| \leq g'' \sqrt{\frac{n(n-1)}{2}}.$$

Proof: Let x be a characteristic vector of unit length corresponding to the characteristic root λ_t; i.e., $Ax = \lambda_t x$ and $(x, x) = 1$. Then

$$\begin{aligned}(Ax, x) &= \lambda_t(x, x) = \lambda_t \\ (A^*x, x) &= (x, Ax) = \bar{\lambda}_t.\end{aligned} \tag{1}$$

Hence

$$Re(\lambda_t) = \frac{(Ax, x) + (A^*x, x)}{2} = \left(\frac{A + A^*}{2} x, x\right) = (Bx, x) \tag{2}$$

and

$$Im(\lambda_t) = \frac{(Ax, x) - (A^*x, x)}{2i} = (Cx, x). \tag{3}$$

Thus

$$\begin{aligned}|\lambda_t| = \left|\sum_{i,j=1}^{n} a_{ij} x_j \bar{x}_i\right| &\leq \sum_{i,j=1}^{n} |a_{ij}| \, |x_i| \, |x_j| \\ &\leq g \sum_{i,j=1}^{n} |x_i| \, |x_j| \\ &= g \left(\sum_{i=1}^{n} |x_i|\right)^2.\end{aligned}$$

Recall the inequality (**II.3.1.2**)

$$\sum_{i=1}^{n} \frac{|x_i|}{n} \leq \left(\sum_{i=1}^{n} \frac{|x_i|^2}{n}\right)^{1/2} \tag{4}$$

which yields

$$g \left(\sum_{i=1}^{n} |x_i|\right)^2 \leq ng \left(\sum_{i=1}^{n} |x_i|^2\right) = ng.$$

Similarly for $Re(\lambda_t)$ and $Im(\lambda_t)$.

Now, if $A + A^t$ is real, then $a_{ij} - \bar{a}_{ji} = -(a_{ji} - \bar{a}_{ij})$. Therefore, by **1.3(3)**,

$$\begin{aligned}|Im(\lambda_t)| &= \left|\sum_{i \neq j} \frac{a_{ij} - \bar{a}_{ji}}{2i} \bar{x}_i x_j\right| \\ &= \left|\sum_{i < j} \frac{a_{ij} - \bar{a}_{ji}}{2} \frac{\bar{x}_i x_j - x_i \bar{x}_j}{i}\right| \\ &\leq \sum_{i < j} \left|\frac{a_{ij} - \bar{a}_{ji}}{2}\right| \left|\frac{\bar{x}_i x_j - x_i \bar{x}_j}{i}\right| \\ &\leq g'' \sum_{i < j} \left|\frac{\bar{x}_i x_j - x_i \bar{x}_j}{i}\right|,\end{aligned}$$

where each $(\bar{x}_i x_j - x_i \bar{x}_j)/i$ is real. Hence, applying the inequality **1.3(4)**,

$$|Im(\lambda_t)| \leq g'' \sqrt{\frac{-n(n-1)}{2} \sum_{i<j} (\overline{x}_i x_j - x_i \overline{x}_j)^2}$$

$$= g'' \sqrt{\frac{n(n-1)}{2}} \sqrt{-\left(\sum_{i=1}^{n} x_i^2 \sum_{i=1}^{n} \overline{x}_i^2 - \left(\sum_{i=1}^{n} x_i \overline{x}_i \right)^2 \right)};$$

this last step follows from Lagrange's identity (**II.3.5.3**). We have

$$|Im(\lambda_t)| \leq g'' \sqrt{\frac{n(n-1)}{2}} \sqrt{\left(\sum_{i=1}^{n} x_i \overline{x}_i \right)^2}$$

$$= g'' \sqrt{\frac{n(n-1)}{2}},$$

where the inequality is strict unless $g'' = 0$.

1.3.2 If $A \in M_n(\mathbf{C})$ has characteristic roots λ_t, $(t = 1, \cdots, n)$, then

$$\mu_n \leq Re(\lambda_t) \leq \mu_1.$$

1.3.3 If A is the matrix in **1.3.2**, then

$$\nu_n \leq Im(\lambda_t) \leq \nu_1.$$

The last two results are proved by a direct application of **II.4.1.5** to **1.3(2)** and **1.3(3)**, since B and C are both hermitian.

1.4 Schur's inequality (1909)

1.4.1 If $A = (a_{ij}) \in M_n(\mathbf{C})$ has characteristic roots λ_t, $(t = 1, \cdots, n)$, then

$$\sum_{t=1}^{n} |\lambda_t|^2 \leq \sum_{i,j=1}^{n} |a_{ij}|^2$$

with equality if and only if A is normal.

Proof: There exists a unitary matrix U such that $U^*AU = T$ where T is an upper triangular matrix (see **I.4.10.2**). It follows that $U^*A^*U = T^*$. Therefore $U^*AA^*U = TT^*$ and tr $(U^*AA^*U) = $ tr (TT^*). Since U^*AA^*U and AA^* are similar, we have

$$\text{tr } (AA^*) = \text{tr } (TT^*). \tag{1}$$

Now, tr $(XX^*) = \sum_{i,j=1}^{n} |x_{ij}|^2$ for any matrix $X = (x_{ij})$ (see **I.2.8.2**). Hence **1.4(1)** gives $\sum_{i,j=1}^{n} |a_{ij}|^2 = \sum_{t=1}^{n} |\lambda_t|^2 + \sum_{i<j} |t_{ij}|^2$. Thus $\sum_{t=1}^{n} |\lambda_t|^2 \leq \sum_{i,j=1}^{n} |a_{ij}|^2$ with equality if and only if $\sum_{i<j} |t_{ij}|^2 = 0$; i.e., if and only if A is normal (see **I.4.10.2**).

1.4.2 Schur's method yields rather simple and elegant proofs of **1.3.1** and **1.2.1**. From the proof of **1.4.1** we get $U^*(A \pm A^*)U/2 = (T \pm T^*)/2$; i.e., $(T + T^*)/2 = U^*BU$ and $(T - T^*)/2 = U^*CU$. Therefore

$$\sum_{t=1}^{n} \left| \frac{\lambda_t + \bar{\lambda}_t}{2} \right|^2 + \sum_{i<j} \frac{|t_{ij}|^2}{2} = \sum_{i,j=1}^{n} |b_{ij}|^2$$

and

$$\sum_{t=1}^{n} \left| \frac{\lambda_t - \bar{\lambda}_t}{2} \right|^2 + \sum_{i<j} \frac{|t_{ij}|^2}{2} = \sum_{i,j=1}^{n} |c_{ij}|^2.$$

Thus

$$\sum_{t=1}^{n} |Re(\lambda_t)|^2 \leq \sum_{i,j=1}^{n} |b_{ij}|^2, \tag{2}$$

$$\sum_{t=1}^{n} |Im(\lambda_t)|^2 \leq \sum_{i,j=1}^{n} |c_{ij}|^2. \tag{3}$$

From 1.4.1, 1.4(2), and 1.4(3) it follows that

$$\sum_{t=1}^{n} |\lambda_t|^2 \leq n^2 g^2, \qquad \sum_{t=1}^{n} |Re(\lambda_t)|^2 \leq n^2 g'^2, \qquad \sum_{t=1}^{n} |Im(\lambda_t)|^2 \leq n^2 g''^2$$

and therefore a fortiori

$$|\lambda_t| \leq ng, \qquad |Re(\lambda_t)| \leq ng', \qquad |Im(\lambda_t)| \leq ng''.$$

If the a_{ij} are real, **1.4(3)** gives

$$\sum_{t=1}^{n} |Im(\lambda_t)|^2 \leq n(n - 1)g''^2.$$

But the $|Im(\lambda_t)|$ occur in pairs, since the coefficients of the characteristic equation of A are real. Therefore

$$|Im(\lambda_t)|^2 \leq \frac{g''^2 n(n - 1)}{2}$$

which is precisely **1.2.1**.

1.5 Browne's theorem

The results discussed so far were obtained in the first decade of this century and most of them give bounds for characteristic roots in terms of the entries or characteristic roots of the related hermitian matrices $(A + A^*)/2$, $(A - A^*)/2i$. There is another important hermitian matrix related to A, viz. A^*A. Recall that the nonnegative square roots of the characteristic roots of A^*A are called the singular values of A (see **I.4.12.2**). Alternatively, singular values of A can be defined as the characteristic roots of the positive semidefinite hermitian factor in the polar factorization of A (see **I.4.19.4**). E. T. Browne obtained in 1928 a result that relates the characteristic roots of A to its singular values.

1.5.1 If $A \in M_n(\mathbf{C})$ has characteristic roots λ_t, $(t = 1, \cdots, n)$, and singular values $\alpha_1 \geq \cdots \geq \alpha_n$, then

$$\alpha_n \leq |\lambda_t| \leq \alpha_1.$$

Proof: Let x be a characteristic vector of unit length corresponding to λ_t. Then $Ax = \lambda_t x$. Therefore $(A*Ax, x) = (Ax, Ax) = \lambda_t \bar{\lambda}_t (x, x) = |\lambda_t|^2$. Now $A*A$ is hermitian and its greatest and least characteristic roots are α_1^2 and α_n^2. Hence, by **II.4.1.5**, $\alpha_n^2 \leq |\lambda_t|^2 \leq \alpha_1^2$.

Recall that the characteristic roots and the singular values of $C_k(A)$, the kth compound matrix of A, are the products taken k at a time of the characteristic roots and singular values, respectively, of A (see **I.2.15.12**).

A direct application of Browne's theorem to $C_k(A)$ yields the following interesting generalization:

$$\prod_{i=1}^{k} \alpha_{n-i+1} \leq \prod_{s=1}^{k} |\lambda_{t_s}| \leq \prod_{i=1}^{k} \alpha_i, \qquad [(t_1, \cdots, t_k) \in Q_{k,n}; k = 1, \cdots, n].$$

The upper bound was obtained previously in **II.4.1.9**.

1.6   Perron's theorem

1.6.1 We have not mentioned yet two other results obtained in the first decade of this century. The first is due to Georg Pick: If λ_t is a characteristic root of a real n-square matrix A, then $|Im(\lambda_t)| \leq g'' \cot (\pi/2n)$.

1.6.2 The other result is the classical theorem of Frobenius. Let R_i and T_j denote the sum of absolute values of the entries of A in the ith row and the sum of the absolute values of the entries in the jth column respectively. Let $R = \max_i R_i$, $T = \max_j T_j$. If all the entries of A are nonnegative, then

$$|\lambda_i| \leq \min (R, T).$$

We shall prove this in a more general form in **2.2.2**.

1.6.3 The first result along similar lines for a general complex matrix was obtained by E. T. Browne in 1930:

$$|\lambda_t| \leq \frac{(R + T)}{2}.$$

1.6.4 The inequality in **1.6.3** was sharpened by W. V. Parker who in 1937 obtained

$$|\lambda_t| \leq \max_i \frac{(R_i + T_i)}{2}, \tag{1}$$

and by A. B. Farnell who in 1944 obtained the following result

$$|\lambda_t| \leq (RT)^{1/2}. \tag{2}$$

Question: Does the inequality

$$|\lambda_t| \leq \max_i (R_i T_i)^{1/2} \tag{3}$$

hold? It turns out that the answer is in the affirmative. In fact **1.6(3)** is a direct corollary to a theorem of Ostrowski [see **2.5(5)**].

1.6.5   In 1946 Alfred Brauer proved that for an arbitrary complex matrix

$$|\lambda_t| \leq \min (R, T).$$

As Brauer acknowledges in a remark added in proof, his result (and incidentally that of Farnell) was anticipated by Oskar Perron in 1933.

1.6.6 (Perron's theorem)   If $c_1, \cdots, c_n$ are any positive numbers and K is the greatest of the n numbers,

$$K_r = \frac{c_1|a_{r1}| + \cdots + c_n|a_{rn}|}{c_r}, \qquad (r = 1, \cdots, n),$$

then $|\lambda_t| \leq K$. For $c_1 = \cdots = c_n = 1$, **1.6.5** is obtained. On the other hand, if **1.6.5** is applied to the matrix $C^{-1}AC$, where C is diag $(c_1, \cdots, c_n)$, Perron's result follows.

1.7 Schneider's theorem

Let $A = (a_{ij}) \in M_n(\mathbf{C})$ and $X = $ diag $(x_1, \cdots, x_n)$, $(x_i > 0; i = 1, \cdots, n)$. Let $r_i = \sum_{j=1}^n |a_{ij}|x_j/x_i$; i.e., $r = (r_1, \cdots, r_n) = X^{-1}CXe$ where

$$e = (1, 1, \cdots, 1) \in \mathbf{C}^n.$$

Let $i_1, \cdots, i_n$ be a permutation of $1, \cdots, n$ such that $r_{i_1} \geq r_{i_2} \geq \cdots \geq r_{i_n}$. Then

$$\prod_{t=1}^k |\lambda_t| \leq \prod_{t=1}^k r_{i_t}, \qquad (k = 1, \cdots, n).$$

2 Regions Containing Characteristic Roots of a General Matrix

2.1 Lévy-Desplanques theorem

2.1.1   Brauer established **1.6.5** by first proving a result (see **2.2**) which is of considerable interest in itself. Unfortunately, in this theorem Brauer was also anticipated, though obviously he was unaware of it, by S. A. Gerŝgorin who in 1931 published the same result. Gerŝgorin proved his theorem as a corollary to the Lévy-Desplanques theorem which in its original form—for certain real matrices—was obtained by Lévy in 1881, was generalized by Desplanques in 1887, and is usually referred to as the "Hadamard theorem" since it appeared in a book by Hadamard in 1903. This result had a most remarkable history. Before it was stated by Hadamard it was also obtained in 1900 by Minkowski with the same restrictions as Lévy's result, and in that form it is still known as the Minkowski theorem. Since then it made regular appearances in the literature until 1949 when an article by Olga Taussky seems to have succeeded in stopping these periodic rediscoveries of the theorem. Incidentally, Gerŝgorin's generalized version of the Lévy-Desplanques theorem is false. The correct generalization along the same lines was stated and proved by Olga Taussky (see 2.2.4).

2.1.2   We introduce the following notation: $P_i = R_i - |a_{ii}|, Q_j = T_j - |a_{jj}|$. Most of the results in this section will be stated in terms of P_i or R_i. Clearly analogous theorems hold for Q_j or T_j.

(Lévy-Desplanques theorem)   If $A = (a_{ij})$ is a complex n-square matrix and

$$|a_{ii}| > P_i, \qquad (i = 1, \cdots, n), \tag{1}$$

then $d(A) \neq 0$.

Proof: Suppose $d(A) = 0$. Then the system $Ax = 0$ has a nontrivial solution $x = (x_1, \cdots, x_n)$ (see **I.3.1.5**).

Let r be the integer for which

$$|x_r| \geq |x_i|, \qquad (i = 1, \cdots, n).$$

Then

$$|a_{rr}| \, |x_r| = \Big| - \sum_{j \neq r} a_{rj} x_j \Big| \leq \sum_{j \neq r} |a_{rj}| \, |x_j| \leq |x_r| P_r$$

which contradicts **2.1**(1).

2.2   Gerŝgorin discs

2.2.1   (Gerŝgorin's theorem)   The characteristic roots of an n-square complex matrix A lie in the closed region of the z-plane consisting of all the discs

$$|z - a_{ii}| \leq P_i, \qquad (i = 1, \cdots, n). \tag{1}$$

Proof: Let λ_t be a characteristic root of A. Then $d(\lambda_t I_n - A) = 0$ and, by the Lévy-Desplanques theorem, $|\lambda_t - a_{ii}| \leq P_i$ for at least one i.

2.2.2   The absolute value of each characteristic root λ_t of A is less than or equal to min (R, T).

For, by **2.2.1**, $|\lambda_t - a_{ii}| \leq P_i$ for some i. Therefore $|\lambda_t| \leq P_i + |a_{ii}| = R_i \leq R$. Similarly, $|\lambda_t| \leq T$.

2.2.3
$$|\lambda_t| \geq k = \min_i \left(|a_{ii}| - P_i\right),$$

$$|d(A)| \leq \min \left(R^n, T^n\right)$$

and, if $k > 0$,
$$|d(A)| \geq k^n.$$

These are immediate consequences of **2.2.1**.

We quote now without proof some related results.

2.2.4 Call an n-square complex matrix indecomposable if it cannot be brought by a simultaneous row and column permutation (see **II.5.2**) to the form $\begin{pmatrix} X & Y \\ 0 & Z \end{pmatrix}$, where X and Z are square submatrices and 0 is a zero submatrix.

If $A = (a_{ij})$ is an indecomposable matrix for which
$$|a_{ii}| \geq P_i, \qquad (i = 1, \cdots, n),$$
with strict inequality for at least one i, then $d(A) \neq 0$.

2.2.5 If $H_{(m)}$ is a set in the complex plane containing m Gerŝgorin discs **2.2(1)** of a matrix A and $H_{(m)}$ has no points in common with the remaining $n - m$ discs, then $H_{(m)}$ contains exactly m characteristic roots of A.

It follows that if A is a real matrix (or merely a complex matrix with real main diagonal entries and real coefficients in its characteristic polynomial) and the Gerŝgorin discs of A are disconnected, then all characteristic roots of A are real.

2.2.6 If A is an indecomposable complex matrix, then all the characteristic roots of A lie inside the union of the discs **2.2(1)** unless a characteristic root is a common boundary point of all n discs.

2.2.7 Let $A = (a_{ij})$ be an n-square complex matrix and let K and k be two positive numbers, $K > k$, such that $|a_{ij}| \leq k$ for $j < i$ and $|a_{ij}| \leq K$ for $j > i$. Then all of the characteristic roots of A lie in the union of the discs
$$|z - a_{ii}| \leq \frac{Kk^{1/n} - kK^{1/n}}{K^{1/n} - k^{1/n}}, \qquad (i = 1, \cdots, n).$$

2.2.8 If λ_t is a characteristic root of A with geometric multiplicity m (see **I.3.12**), then λ_t lies in at least m Gerŝgorin discs **2.2(1)**.

2.2.9 Let A be an n-square complex matrix and let σ be a permutation on $1, \cdots, n$. Then the characteristic roots of A lie in the union of the n regions
$$|z - a_{ii}| \leq P_i, \quad \text{if } i = \sigma(i),$$
$$|z - a_{ii}| \geq L_i, \quad \text{if } i \neq \sigma(i)$$
where $L_i = |a_{i\sigma(i)}| - \sum_{\substack{j \neq i \\ j \neq \sigma(i)}} |a_{ij}|$.

2.2.10 Let A, σ, P_i, and L_i be defined as in **2.2.9**. Suppose that λ_t is a characteristic root of A with geometric multiplicity m (see **I.3.12**). Then λ_t lies in at least m of the regions

$$|z - a_{ii}| \le P_i, \quad \text{if } i = \sigma(i),$$
$$|z - a_{ii}| \ge L_i, \quad \text{if } i \ne \sigma(i).$$

2.2.11 Let λ_t be a characteristic root of $A = (a_{ij}) \in M_n(\mathbf{C})$ with geometric multiplicity m. If $\beta_1, \cdots, \beta_n$ are positive numbers such that $\sum\limits_{i=1}^{n} 1/(1 + \beta_i) \le m$, then λ_t lies in at least one of the discs

$$|z - a_{ii}| \le \beta_i P_i, \qquad (i = 1, \cdots, n).$$

2.2.12 If $|a_{ii}| > P_i$ for at least r subscripts i, then the rank of A is at least r.

2.2.13 If the rank of $A = (a_{ij}) \in M_n(\mathbf{C})$ is $\rho(A)$, then

$$\sum_{i=1}^{n} \frac{|a_{ii}|}{R_i} \le \rho(A) \tag{2}$$

where $R_i = \sum\limits_{j=1}^{n} |a_{ij}|$. We agree that $0/0$, if it occurs on the left-hand side of **2.2(2)**, is to be interpreted as 0.

2.3 Example

The characteristic roots of

$$A = \begin{pmatrix} 7 + 3i & -4 - 6i & -4 \\ -1 - 6i & 7 & -2 - 6i \\ 2 & 4 - 6i & 13 - 3i \end{pmatrix}$$

are 9, $9 + 9i$, and $9 - 9i$; their absolute values are 9 and 12.73. Hirsch's theorem **1.3.1** yields

$$|\lambda_t| \le 40.03, \qquad |Re(\lambda_t)| \le 39, \qquad |Im(\lambda_t)| \le 20.12.$$

Schur's inequality **1.4.1** reads $\sum\limits_{i=1}^{3} |\lambda_t|^2 = 405 \le \sum\limits_{i,j=1}^{3} |a_{ij}|^2 = 486$; while the theorems of Browne (**1.6.3**), Parker [**1.6(1)**], Farnell [**1.6(2)**], and Perron (Brauer) (**1.6.5**) give in order the following upper bounds for $|\lambda_t|$: 23.11, 23.11, 23.10, 22.55. The Gerŝgorin discs

$$|z - 7 - 3i| \le 11.21, \ |z - 13 + 3i| \le 9.21$$

contain two characteristic roots each.

2.4   Ovals of Cassini

2.4.1   If $A = (a_{ij})$ is a complex n-square matrix and

$$|a_{ii}| \, |a_{jj}| > P_i P_j, \qquad (i, j = 1, \cdots, n; i \neq j), \tag{1}$$

then $d(A) \neq 0$.

Proof: Suppose that the inequalities **2.4**(1) hold and $d(A) = 0$. Then the system $Ax = 0$ has a nontrivial solution $x = (x_1, \cdots, x_n)$ (see **I.3.1.5**). Let x_r and x_s be coordinates of x such that

$$|x_r| \geq |x_s| \geq |x_i|, \qquad (i = 1, \cdots, r - 1, r + 1, \cdots, n).$$

Note that $x_s \neq 0$. For, if $x_s = 0$, then $x_i = 0$ for all $i \neq r$, and $x_r \neq 0$. But $Ax = 0$ and, in particular, $\sum_{j=1}^{n} a_{rj}x_j = 0$; i.e., $a_{rr}x_r = 0$. Thus $a_{rr} = 0$ which contravenes **2.4**(1). Hence $x_s \neq 0$. The supposition $d(A) = 0$ implies therefore that

$$|a_{rr}| \, |x_r| = \Big| - \sum_{j \neq r} a_{rj}x_j \Big| \leq \sum_{j \neq r} |a_{rj}| \, |x_j| \leq |x_s| P_r,$$

$$|a_{ss}| \, |x_s| = \Big| - \sum_{j \neq s} a_{sj}x_j \Big| \leq \sum_{j \neq s} |a_{sj}| \, |x_j| \leq |x_r| P_s.$$

and $|a_{rr}| \, |a_{ss}| \leq P_r P_s$, contradicting **2.4**(1).

2.4.2   Each characteristic root of $A = (a_{ij}) \in M_n(\mathbf{C})$ lies in at least one of the $n(n-1)/2$ ovals of Cassini

$$|z - a_{ii}| \, |z - a_{jj}| \leq P_i P_j, \qquad (i, j = 1, 2, \cdots, n; i \neq j).$$

Proof: Let λ_t be a characteristic root of A. Then $d(\lambda_t I_n - A) = 0$ and, by **2.4.1**, $|\lambda_t - a_{ii}| \, |\lambda_t - a_{jj}| > P_i P_j$ cannot hold for all i, j, $(i \neq j)$.

2.4.3   The following question arises naturally: For what values of k must each characteristic root of A lie within at least one of the $\binom{n}{k}$ regions

$$\prod_{\nu=1}^{k} |z - a_{i_\nu i_\nu}| \leq \prod_{\nu=1}^{k} P_{i_\nu}?$$

The answer is, rather disappointingly, $k \leq 2$. We are indebted to Dr. Morris Newman for the following example: If

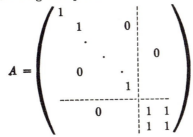

and $k > 2$, all regions collapse into the point $z = 1$ which certainly does not contain the characteristic roots 2 and 0.

2.5 Ostrowski's theorems

2.5.1 Let $0 \leq \alpha \leq 1$, $A \in M_n(\mathbf{C})$ and let P_i and Q_j be defined as in **2.1.2**. Then $d(A) \neq 0$ provided

$$|a_{ii}| > P_i^\alpha Q_i^{1-\alpha}, \qquad (i = 1, \cdots, n). \tag{1}$$

Proof: Assume that $0 < \alpha < 1$, since $\alpha = 0$ or 1 is the Lévy-Desplanques theorem. (Alternatively, the Lévy-Desplanques theorem can be obtained from the $0 < \alpha < 1$ case by the usual limiting process.) We assume also that $P_i > 0$ for all i. For, if $P_i = 0$, all off-diagonal entries in the ith row are 0 and the problem is essentially reduced to the $(n - 1)$-square case.

We suppose then that the inequalities **2.5(1)** hold, that $d(A) = 0$, and proceed to deduce a contradiction. The fact that $d(A) = 0$ implies that $Ax = 0$ has a nontrivial solution $x = (x_1, \cdots, x_n)$ (see **I.3.1.5**); i.e., $\sum_{j=1}^{n} a_{ij} x_j = 0$, $(i = 1, \cdots, n)$. Therefore $|a_{ii}| |x_i| \leq \sum_{j \neq i} |a_{ij}| |x_j|$ and, since $|a_{ii}| > P_i^\alpha Q_i^{1-\alpha}$,

$$P_i^\alpha Q_i^{1-\alpha} |x_i| \leq \sum_{j \neq i} |a_{ij}|^\alpha |a_{ij}|^{1-\alpha} |x_j|, \qquad (i = 1, \cdots, n), \tag{2}$$

with strict inequality for at least one i, viz. for the i for which $x_i \neq 0$. Recall the Hölder inequality in the following form (see **II.3.3**): if $(y_1, \cdots, y_n)$ and $(z_1, \cdots, z_n)$ are nonnegative n-tuples and $0 < \alpha < 1$, then

$$\sum_{i=1}^{n} y_i z_i \leq \left(\sum_{i=1}^{n} y_i^{1/\alpha} \right)^\alpha \left(\sum_{i=1}^{n} z_i^{1/(1-\alpha)} \right)^{1-\alpha}.$$

Applying this inequality to the right hand side of **2.5(2)** we obtain

$$P_i^\alpha Q_i^{1-\alpha} |x_i| \leq \left(\sum_{j \neq i} |a_{ij}^\alpha|^{1/\alpha} \right)^\alpha \left(\sum_{j \neq i} |a_{ij}^{1-\alpha} x_j|^{1/(1-\alpha)} \right)^{1-\alpha}$$

$$= P_i^\alpha \left(\sum_{j \neq i} |a_{ij}| |x_j|^{1/(1-\alpha)} \right)^{1-\alpha}.$$

Therefore

$$Q_i |x_i|^{1/(1-\alpha)} \leq \sum_{j \neq i} |a_{ij}| |x_j|^{1/(1-\alpha)}, \qquad (i = 1, \cdots, n).$$

Now sum up with respect to i, remembering that the inequality is strict for at least one i:

$$\sum_{i=1}^{n} Q_i |x_i|^{1/(1-\alpha)} < \sum_{\substack{i,j \\ i \neq j}} |a_{ij}| |x_j|^{1/(1-\alpha)}$$

$$= \sum_{j=1}^{n} \left(|x_j|^{1/(1-\alpha)} \sum_{\substack{i=1 \\ i \neq j}}^{n} |a_{ij}| \right)$$

$$= \sum_{j=1}^{n} |x_j|^{1/(1-\alpha)} Q_j,$$

a contradiction.

2.5.2 $d(A) \neq 0$, if $|a_{ii}| > \alpha P_i + (1 - \alpha)Q_i$, $(i = 1, \cdots, n)$. This follows directly from **II.3.3(3)** and **2.5.1**.

2.5.3   Each characteristic root of $A = (a_{ij})$ lies on or inside at least one of the discs

$$|z - a_{ii}| \leq P_i^\alpha Q_i^{1-\alpha}, \qquad (i = 1, \cdots, n).$$

Proof: If λ_t is a characteristic root of A, then $d(\lambda_t I_n - A) = 0$ and the result follows immediately from **2.5.1**.

2.5.4 $$|\lambda_t| \leq \max_i (|a_{ii}| + P_i^\alpha Q_i^{1-\alpha}). \tag{3}$$

For, by **2.5.3**, $P_i^\alpha Q_i^{1-\alpha} \geq |\lambda_t - a_{ii}| \geq |\lambda_t| - |a_{ii}|$ for some i. Applying the Hölder inequality [see **II.2.2(1)**] to **2.5(3)** we obtain

$$|\lambda_t| \leq \max_i (|a_{ii}| + P_i)^\alpha (|a_{ii}| + Q_i)^{1-\alpha} = \max_i (R_i^\alpha T_i^{1-\alpha}). \tag{4}$$

For $\alpha = \frac{1}{2}$ we have

$$|\lambda_t| \leq \max_i (R_i T_i)^{1/2}. \tag{5}$$

2.5.5 $|\lambda_t| \leq \max_i (|a_{ii}| + \alpha P_i + (1 - \alpha)Q_i)$; $\alpha = \frac{1}{2}$ gives Parker's theorem [see **1.6(1)**]; $\alpha = 0$ or 1 gives **1.6.5**.

2.5.6   If $0 \leq \alpha \leq 1$ and for all pairs of subscripts i, j, $i \neq j$,

$$|a_{ii}| |a_{jj}| > P_i^\alpha Q_i^{1-\alpha} P_j^\alpha Q_j^{1-\alpha}, \tag{6}$$

then $d(A) \neq 0$.

Proof: Let $s_i = P_i^\alpha Q_i^{1-\alpha}/|a_{ii}|$ and let $s_{k_1} \geq s_{k_2} \geq \cdots \geq s_{k_n}$. Then **2.5(1)** is equivalent to $s_i < 1$ while **2.5(6)** is equivalent to $s_i s_j < 1$. In particular, $s_{k_1} s_{k_2} < 1$ and thus $s_{k_2} < 1$. If $s_{k_1} = 0$, then $s_{k_i} = 0$, $(i = 2, \cdots, n)$, and $d(A) = a_{11} a_{22} \cdots a_{nn} \neq 0$. Otherwise $0 < s_{k_2} < 1$. Let $q = \sqrt{s_{k_1}/s_{k_2}}$. Let B be the matrix obtained from A by multiplying the k_1th row and the k_1th column by q. For B use the same notation as for A but with primes. Then $P_{k_1}' = qP_{k_1}$, $Q_{k_1}' = qQ_{k_1}$, $|a_{k_1 k_1}'| = q^2|a_{k_1 k_1}|$ and therefore $s_{k_1}' = s_{k_1}/q = \sqrt{s_{k_1} s_{k_2}} < 1$. Further, for $i > 1$, $s_{k_i}' \leq qs_{k_i} \leq qs_{k_2} = \sqrt{s_{k_1} s_{k_2}} < 1$. Therefore, by **2.5.1**, $d(B) \neq 0$ and, since $d(B) = q^2 d(A)$, $d(A) \neq 0$.

2.5.7   As an immediate consequence of **2.5.6** we obtain the following result which generalizes **2.2.2**.

Each characteristic root of $A = (a_{ij}) \in M_n(\mathbf{C})$ lies on or inside at least one of the $n(n - 1)/2$ ovals of Cassini;

$$|z - a_{ii}| |z - a_{jj}| \leq (P_i P_j)^\alpha (Q_i Q_j)^{1-\alpha}, \qquad (i, j = 1, \cdots, n; i \neq j; 0 \leq \alpha \leq 1).$$

For, if $|\lambda_t - a_{ii}| |\lambda_t - a_{jj}| > (P_i P_j)^\alpha (Q_i Q_j)^{1-\alpha}$ for all i, j, $(i \neq j)$, then, by **2.5.6**, $d(\lambda_t I_n - A) \neq 0$.

2.6 Fan's theorem

In conclusion we state and prove an elegant theorem due to Ky Fan generalizing a result by Kotelyanskiĭ.

If $A = (a_{ij}) \in M_n(\mathbf{C})$ and $B = (b_{ij}) \in M_n(\mathbf{R})$ is a nonnegative matrix such that $|a_{ij}| \leq b_{ij}$, $(i, j = 1, \cdots, n; i \neq j)$, then each characteristic root of A lies in at least one of the n circular discs $|z - a_{ii}| \leq \mu - b_{ii}$, where μ denotes the maximal characteristic root of B (see **II.5.7.1**).

Proof: Assume that all b_{ij} are positive, since the result can be extended to the general case by a continuity argument. Now, by the Perron-Frobenius theorem (see **II.5.5.1**), B has a positive characteristic vector $(x_1, \cdots, x_n)$ corresponding to μ: $\sum_{j=1}^{n} b_{ij}x_j = \mu x_i$, $(i = 1, \cdots, n)$. Therefore,

$$\frac{1}{x_i} \sum_{j \neq i} |a_{ij}| x_j \leq \frac{1}{x_i} \sum_{j \neq i} b_{ij}x_j$$

$$= \frac{1}{x_i} \sum_{j=1}^{n} b_{ij}x_j - \frac{1}{x_i} b_{ii}x_i$$

$$= \frac{1}{x_i} \mu x_i - b_{ii}$$

$$= \mu - b_{ii}.$$

But, by the Gerŝgorin theorem (see **2.2.1**) applied to the matrix $X^{-1}AX$, where $X = \mathrm{diag}\,(x_1, \cdots, x_n)$,

$$|\lambda - a_{ii}| \leq \frac{1}{x_i} \sum_{j \neq i} a_{ij}x_j, \text{ for some } i.$$

3   Characteristic Roots of Special Types of Matrices

3.1   Nonnegative matrices

In this section we shall be concerned exclusively with matrices with non-negative entries (see **II.5**).

Let S_i be the sum of the entries in row i of A, let $s = \min_i S_i$, $S = \max_i S_i$, and let r denote the maximal characteristic root of A. We will state the following theorems in terms of row sums. In view of **I.2.15.13** analogous theorems clearly hold for column sums.

3.1.1   (Frobenius theorem) $s \leq r \leq S$.

Proof: We can assume that $A > 0$ by the standard continuity argument. Let $x = (x_1, \cdots, x_n)$ be a positive characteristic vector corresponding to r. Let $x_M = \max\limits_i (x_i)$ and $x_\mu = \min\limits_i (x_i)$. Then

$$\sum_{j=1}^{n} a_{ij}x_j = rx_i, \qquad (i = 1, \cdots, n), \tag{1}$$

and in particular

$$rx_M = \sum_{j=1}^{n} a_{Mj}x_j \le \sum_{j=1}^{n} a_{Mj}x_M = x_M S_M. \tag{2}$$

Since $x_M > 0$, $r \le S_M$ and therefore $r \le S$. Similarly, $r \ge s$.

Note that this result is the best possible. In fact if $S = s$, i.e., if the matrix is a scalar multiple of a row stochastic matrix (see **II.5.13**), then $r = S = s$. Conversely, if A is positive (this is a necessary condition) and $r = S$, then it follows from **3.1(2)** that all x_i are equal. Hence, by **3.1(1)**, all S_i are equal and $S = s$. We shall assume henceforth that A is a positive n-square matrix.

3.1.2 Ledermann proposed the following interesting problem: If not all S_i are equal, determine positive numbers p_1 and p_2 such that

$$s + p_2 \le r \le S - p_1.$$

He proved the following result: Let

$$m = \min_{i,j} (a_{ij}) \quad \text{and} \quad \delta = \max_{S_i < S_j} (S_i/S_j).$$

Then

$$s + m\left(\frac{1}{\sqrt{\delta}} - 1\right) \le r \le S - m(1 - \sqrt{\delta}). \tag{3}$$

Proof: As in **3.1** let x be the positive characteristic vector. Clearly not all x_i are equal. Let $x_M = \max\limits_i (x_i)$ and $x_\mu = \min\limits_i (x_i)$. Then, by **3.1(2)**, $S_\mu < r < S_M$ and therefore $S_\mu/S_M < 1$. Also

$$rx_\mu = \sum_{j=1}^{n} a_{\mu j}x_j < x_M S_\mu \tag{4}$$

and

$$rx_M = \sum_{j=1}^{n} a_{Mj}x_j > x_\mu S_M. \tag{5}$$

From **3.1(4)** and **3.1(5)** we have

$$\frac{x_\mu}{x_M} < \frac{x_M}{x_\mu} \frac{S_\mu}{S_M}$$

and therefore

$$\frac{x_\mu}{x_M} < \sqrt{\frac{S_\mu}{S_M}} \le \sqrt{\delta}.$$

By **3.1(5)**,

$$r = \sum_{j=1}^{n} \frac{a_{Mj}x_j}{x_M}$$

$$< a_{M1} + a_{M2} + \cdots + a_{M,\mu-1} + a_{M\mu}\sqrt{\delta} + \cdots + a_{Mn}$$

$$= S_M - a_{M\mu}(1 - \sqrt{\delta})$$

$$\leq S - m(1 - \sqrt{\delta}).$$

Similarly,

$$r = \sum_{j=1}^{n} \frac{a_{\mu j}x_j}{x_\mu}$$

$$> a_{\mu 1} + a_{\mu 2} + \cdots + \frac{a_{\mu M}}{\sqrt{\delta}} + \cdots + a_{\mu n}$$

$$= S_\mu + a_{\mu M}\left(\frac{1}{\sqrt{\delta}} - 1\right)$$

$$\geq s + m\left(\frac{1}{\sqrt{\delta}} - 1\right).$$

3.1.3   Ledermann's result was improved by Ostrowski.

If $A > 0$, r, S, s, m are defined as in **3.1.1** and **3.1.2**; and if $s < S$, then

$$s + m\left(\frac{1}{\sigma} - 1\right) \leq r \leq S - m(1 - \sigma) \tag{6}$$

where $\sigma = \sqrt{(s - m)/(S - m)}$.

Proof: Let $(x_1, \cdots, x_n)$ be a positive characteristic vector corresponding to r. For simplicity assume that $1 = x_1 \geq x_2 \geq \cdots \geq x_n$; this can be achieved by premultiplication of A by a permutation matrix and postmultiplication by its inverse. Let k and t be any subscripts:

$$rx_k \geq a_{k1} + \left(\sum_{j=2}^{n} a_{kj}\right)x_n = a_{k1}(1 - x_n) + S_k x_n.$$

Therefore

$$r \geq \frac{x_n S_k + (1 - x_n)m}{x_k}. \tag{7}$$

Similarly,

$$rx_t \leq \sum_{j=1}^{n-1} a_{tj} + a_{tm}x_n = S_t - (1 - x_n)a_{tn}$$

and thus

$$r \leq \frac{S_t - (1 - x_n)m}{x_t}. \tag{8}$$

We now specialize: If $S_k = S$ and $S_t = s$, **3.1(7)** and **3.1(8)** yield

$$r \geq x_n S + (1 - x_n)m = x_n(S - m) + m$$

and $r \leq (s - m)/x_n + m$. Hence

$$x_n(S - m) + m \le r \le \frac{s - m}{x_n} + m,$$

$$x_n(S - m) \le r - m \le \frac{s - m}{x_n},$$

and $x_n \le \sqrt{(s - m)/(S - m)} = \sigma$. Now put $k = n$ and $t = 1$ in **3.1**(7) and **3.1**(8): $S_n + (1/x_n - 1)m \le r \le S_1 - (1 - x_n)m$ and

$$s + (1/\sigma - 1)m \le r \le S - (1 - \sigma)m.$$

Ostrowski's result is sharper than Ledermann's since

$$\frac{s - m}{S - m} \le \frac{s}{S} \le \max_{S_i < S_j} \frac{S_i}{S_j}.$$

3.1.4 Brauer improved these bounds for the maximal root of a positive matrix and actually showed that his is the best possible result involving m, S, and s, in the sense that there exist matrices for which his bounds are attained.

If $A = (a_{ij}) > 0$, and r, s, S, m are defined as in **3.1** and **3.2**, then

$$s + m(h - 1) \le r \le S - m\left(1 - \frac{1}{g}\right) \tag{9}$$

where

$$g = \frac{S - 2m + \sqrt{S^2 - 4m(S - s)}}{2(s - m)}, \qquad h = \frac{-s + 2m + \sqrt{s^2 + 4m(S - s)}}{2m}.$$

Proof: Assume without loss of generality that $S_1 = S$ and $S_n = s$. Let B be the matrix obtained from A by multiplying the entries in the last row of A by g and those in the last column by $1/g$. Obviously, A and B are similar and have the same characteristic roots. The ith row sum of B, $(i = 1, \cdots, n - 1)$, is $\sum_{j=1}^{n} a_{ij} - a_{in} + a_{in}/g = S_i - a_{in}(1 - 1/g) \le S - m(1 - 1/g) = K_1$, say. The nth row sum of B is equal to $\sum_{j=1}^{n} a_{nj}g - a_{nn}g + a_{nn} = gs - a_{nn}(g - 1) \le gs - m(g - 1) = K_2$, say. We prove now that $K_1 = K_2$ with g as defined in the statement of the theorem. For, $S - m(1 - 1/g) = gs - m(g - 1)$ is implied by $Sg - mg + m = g^2s - mg^2 + mg$ and thus by $g^2(s - m) - (S - 2m)g - m = 0$ which in turn is implied by

$$g = \frac{S - 2m + \sqrt{(S - 2m)^2 + 4m(s - m)}}{2(s - m)}.$$

Hence for this value of g all row sums of B are bounded above by $S - m(1 - 1/g)$ and therefore, using **3.1.1** we have

$$r \le S - m(1 - 1/g).$$

In order to obtain the lower bound we construct a matrix C by dividing the entries in the first row of A by h and multiplying those in the first column by h. Then the first row sum of C is

$$\frac{\sum\limits_{j=1}^{n} a_{1j}h}{h + a_{11}} - a_{11} = \frac{S}{h + a_{11}}\left(1 - \frac{1}{h}\right) \geq \frac{S}{h + m}\left(1 - \frac{1}{h}\right) = K_3.$$

The ith row sum of C, $(i = 2, \cdots, n)$, is $\sum\limits_{j=1}^{n} a_{ij} - a_{i1} + a_{i1}h = S_i + a_{i1}(h - 1) \geq s + m(h - 1) = K_4$. A straightforward computation shows that $K_3 = K_4$ when h has the value indicated in the statement of the theorem. Thus all row sums of C are bounded below by $s + m(h - 1)$ and, by **3.1.1**,

$$r \geq s + m(h - 1).$$

We omit the part of Brauer's proof showing that his result is a refinement of Ostrowski's theorem **3.1.3**.

We now exhibit two matrices P and Q for which it is assumed that the numbers $S > s \geq nm > 0$ are prescribed and whose maximal characteristic roots attain the upper and the lower bound in **3.1(9)** respectively. Let

$$P = \begin{pmatrix} P_1 & \vdots & P_2 \\ \text{----} & \vdots & \text{----} \\ P_3 & \vdots & m \end{pmatrix}$$

where every entry is not less than m, the row sums of the $(n - 1)$-square submatrix P_1 are equal to $S - m$, all entries of P_2 are equal to m, while those of P_3 add up to $s - m$. Then the matrix

$$\begin{pmatrix} P_1 & \vdots & \dfrac{1}{g} P_2 \\ \text{----} & \vdots & \text{------} \\ gP_3 & \vdots & m \end{pmatrix}$$

is similar to P. Each of its first $n - 1$ row sums is equal to $S - m + m/g = K_1$; its last row sum is equal to $gs - gm + m = K_2 = K_1$. Hence, by **3.1.1**, the maximal root of P is K_1. Similarly, we construct

$$Q = \begin{pmatrix} m & \vdots & Q_2 \\ \text{----} & \vdots & \text{---} \\ Q_3 & \vdots & Q_4 \end{pmatrix}$$

where every entry is not less than m, the row sums of the $(n - 1)$-square submatrix Q_4 are all equal to $s - m$, all entries of Q_3 are equal to m, while those of Q_2 add up to $S - m$. Now, if the first row is divided by h and the first column multiplied by h the resulting matrix is similar to Q and is a multiple of a row stochastic matrix with row sum $K_3 = K_4$.

Thus Brauer's bounds are the best obtainable using S, s, m. If more information about the matrix is used, better bounds can be obtained.

3.1.5 Let r be the maximal characteristic root of an n-square nonnegative indecomposable (see **II.5.5.1**) matrix $A = (a_{ij})$. Let S_i, s, S be defined as in **3.1**. Let $\sigma = \sum_{i=1}^{n} S_i/n$, $\alpha = \min_i a_{ii}$, and k be the least of the positive off-diagonal elements of A. Then

$$\frac{(n-1)(1-\epsilon)s + n\epsilon\sigma}{(n-1)(1-\epsilon) + n\epsilon} \leq r \leq \frac{(n-1)(1-\epsilon)S + n\epsilon\sigma}{(n-1)(1-\epsilon) + n\epsilon}, \tag{10}$$

where $\epsilon = (k/(S-\alpha))^{n-1}$, so that

$$s + \epsilon(\sigma - s) \leq r \leq S - \epsilon(S - \sigma). \tag{11}$$

We omit the proof of this result.

3.1.6   Let A, B be two indecomposable nonnegative matrices with maximal characteristic roots α and β respectively and such that $m = \min_{i,j} (b_{ij} - a_{ij}) \geq 0$; let $\max_{i,j} (b_{ij} - a_{ij})$ be denoted by M. Let R be any nonnegative matrix which commutes with A or B and has no zero columns. Then

$$\frac{m}{\max_i \max_j \dfrac{r_{ij}}{\sum_{k=1}^{n} r_{kj}}} \leq \beta - \alpha \leq \frac{M}{\min_i \min_j \dfrac{r_{ij}}{\sum_{k=1}^{n} r_{kj}}}.$$

Proof: Let $E^{(n)}$ denote the $(n-1)$-simplex of nonnegative n-tuples whose coordinates add up to 1 (see **II.1.6, II.1.9.1**). Let $x \in E^{(n)}$ be the positive characteristic vector of A corresponding to α. Let R commute with A. Define i as follows:

$$\frac{(Bx)_i}{x_i} = \max_k \frac{(Bx)_k}{x_k}.$$

Then, by **II.5.5.2**, $(Bx)_i/x_i \geq \beta$. We assert that $\beta - \alpha \leq M/x_i$. For, $(B - A)x + Ax = Bx$, $(B - A)x + \alpha x = Bx$, and hence, considering the ith coordinate,

$$\sum_{k=1}^{n} (b_{ik} - a_{ik})x_k + \alpha x_i = (Bx)_i,$$

i.e.,

$$\frac{1}{x_i} \sum_{k=1}^{n} (b_{ik} - a_{ik})x_k + \alpha = \frac{(Bx)_i}{x_i} \geq \beta.$$

Since $x \in E^{(n)}$, $\sum_{k=1}^{n} (b_{ik} - a_{ik})x_k \leq M$ and thus

$$\beta - \alpha \leq \frac{M}{x_i} \leq \frac{M}{\min_k x_k}.$$

We require a lower bound for $\min_k x_k$. Let $\sigma(z)$ denote the sum of coordinates of any vector z. By **II.5.6**, $x = Rx/\sigma(Rx)$. Therefore

$$x_k = \frac{(Rx)_k}{\sigma(Rx)} \geq \min_{y \in E^{(n)}} \frac{(Ry)_k}{\sigma(Ry)}.$$

It is easy to show that

$$\min_{y \in E^{(n)}} \frac{\sum_{j=1}^{n} r_{kj} y_j}{\sum_{j=1}^{n} \sum_{s=1}^{n} r_{sj} y_j} = \min_{t} \frac{r_{kt}}{\sum_{s=1}^{n} r_{st}}.$$

Thus

$$\beta - \alpha \leq \frac{M}{\min_{k} x_k} \leq \frac{M}{\min_{k} \min_{t} \dfrac{r_{kt}}{\sum_{s=1}^{n} r_{st}}}.$$

The remainder of the theorem is proved in a similar way. Note that R is any nonnegative matrix commuting with A or B; e.g., any nonnegative polynomial in A with no zero column. It is not known what the best choices are for R.

3.2 Example

Let

$$A = \begin{pmatrix} 1 & 1 & 2 \\ 2 & 1 & 3 \\ 2 & 3 & 5 \end{pmatrix}.$$

Then,

by **3.1.1**,	$4 \leq r \leq 10$	(rows),	
	$5 \leq r \leq 10$	(columns);	
by **3.1(3)**,	$4.225 \leq r \leq 9.816$	(rows),	
	$4.414 \leq r \leq 9.707$	(columns);	
by **3.1(6)**,	$4.732 \leq r \leq 9.577$	(rows),	
	$4.500 \leq r \leq 9.667$	(columns);	
by **3.1(7)**,	$5.162 \leq r \leq 9.359$	(rows),	
	$4.854 \leq r \leq 9.472$	(columns);	

whereas,

by **3.1(10)**,	$4.049 \leq r \leq 9.939$	(rows),
	$5.031 \leq r \leq 9.939$	(columns).

The actual value of r to 4 significant figures is 7.531.

3.3 Stability matrices

Call a complex matrix *stable* (*semistable*), if the real parts of all its characteristic roots are negative (nonpositive). Such matrices are of paramount importance in the theory of stability of solutions of differential equations.

3.3.1   If $A = (a_{ij}) \in M_n(\mathbf{R})$, $a_{ij} \geq 0$ for all i, j, $(i \neq j)$, and there exist positive numbers $t_1, \cdots, t_n$ such that

$$\sum_{j=1}^{n} t_j a_{ij} \leq 0, \qquad (i = 1, \cdots, n), \tag{1}$$

then A is semistable.

Proof: Let λ_t be a characteristic root of A and $(x_1, \cdots, x_n)$ a characteristic vector corresponding to λ_t. Let $y_i = x_i/t_i$, $(i = 1, \cdots, n)$, and $|y_m| = \max_i |y_i|$.

Then $\lambda_t t_i y_i = \sum_{j=1}^{n} a_{ij} t_j y_j$ and

$$\lambda_t t_m = t_m a_{mm} + \frac{\sum\limits_{j \neq m} t_j a_{mj} y_j}{y_m}.$$

Therefore

$$|\lambda_t t_m - t_m a_{mm}| \leq \sum_{j \neq m} |t_j a_{mj}| \left| \frac{y_j}{y_m} \right|$$

$$\leq \sum_{j \neq m} t_j a_{mj}$$

$$\leq -t_m a_{mm}, \qquad \text{by } \mathbf{3.3(1)}.$$

Thus $|\lambda_t - a_{mm}| \leq -a_{mm}$ and in the complex plane the characteristic root λ_t lies in the disc which passes through 0 and whose center is at the negative number a_{mm}.

3.3.2   If the inequalities **3.3(1)** are strict, then A is stable.

3.3.3   Rohrbach generalized **3.3.1** to complex matrices; the inequalities **3.3(1)** are replaced in his theorem by

$$\sum_{j \neq i} t_j |a_{ij}| \leq -t_i Re(a_{ii}), \qquad (i = 1, \cdots, n).$$

3.3.4   If A is defined as in **3.3.1** and $\epsilon = \max_i (-a_{ii})$, then $|\lambda_t + \epsilon| \leq \epsilon$.

This corollary to **3.3.1** was deduced by Fréchet who used it to prove his theorem on characteristic roots of row stochastic matrices (see **3.4.1**).

3.3.5   If $A = (a_{ij}) \in M_n(\mathbf{C})$ and

$$Re(a_{ii}) \leq -\sum_{\substack{j=1 \\ j \neq i}}^{n} |a_{ij}|, \qquad (i = 1, \cdots, n), \tag{2}$$

then A is semistable. If the inequalities in **3.3(2)** are strict, then A is stable. If A is real and satisfies **3.3(2)**, then A is stable if and only if it is nonsingular. These results follow directly from **2.2.1**.

3.3.6 Let $A = (a_{ij}) \in M_n(\mathbf{R})$, $(a_{ii} \leq 0$, $a_{ij} \geq 0$ for $i \neq j)$. Then A is stable if and only if there exist positive numbers $x_1, \cdots, x_n$ such that $X^{-1}AXe < 0$ where $X = \mathrm{diag}\,(x_1, \cdots, x_n)$ and $e = (1, \cdots, 1) \in \mathbf{R}^n$. This result clearly implies **3.3.2**.

3.3.7 (Lyapunov's theorem) The only known general result on stability of matrices is due to Lyapunov. It can be stated as follows.

A matrix $A \in M_n(\mathbf{C})$ is stable, if and only if there exists a negative definite (see **I.4.12**) matrix H such that $AH + HA^*$ is positive definite.

3.3.8 The result **3.3.7** can be proved from the following result by Ostrowski and Schneider.

Let $A \in M_n(\mathbf{C})$. Then A has no purely imaginary characteristic root, if and only if there exists a hermitian matrix H such that $AH + HA^*$ is positive definite; and then A and H have the same number of characteristic roots with negative real parts.

3.3.9 A result of Taussky that also is a corollary of **3.3.8** follows.

Let $A \in M_n(\mathbf{C})$ and suppose that α_i, $(i = 1, \cdots, n)$, are the characteristic roots of A and that they satisfy $\prod_{i,k=1}^{n} (\alpha_i + \bar{\alpha}_k) \neq 0$. Then there exists a unique nonsingular hermitian matrix H satisfying $AH + HA^* = I_n$. Moreover H has as many positive characteristic roots as there are α_i with positive real parts.

3.3.10 The equation $AH + HA^* = I_n$ is precisely the one considered in **I.1.10**. Let U be a unitary matrix in $M_n(\mathbf{C})$ for which $UAU^* = T$ is upper triangular with the characteristic roots $\alpha_1, \cdots, \alpha_n$ appearing on the main diagonal of T. Then $AH + HA^* = I_n$ becomes $TK + KT^* = I_n$ where $K = UHU^*$. This last system is equivalent to the equation $(I_n \otimes T + \bar{T} \otimes I_n)x = e$ (see **I.1.10**). The matrix $I_n \otimes T + \bar{T} \otimes I_n \in M_{n^2}(\mathbf{C})$ is triangular with $\alpha_i + \bar{\alpha}_k$, $(1 \leq i, k \leq n)$, as main diagonal elements. Thus a unique solution x exists if and only if $d(I_n \otimes T + \bar{T} \otimes I_n) = \prod_{i,k=1}^{n} (\alpha_i + \bar{\alpha}_k) \neq 0$. This accounts for the hypothesis in **3.3.9**.

3.4 Row stochastic matrices

Recall that $P = (p_{ij}) \in M_n(\mathbf{R})$ is row stochastic if

$$p_{ij} \geq 0, \; (i, j = 1, \cdots, n), \quad \text{and} \quad \sum_{j=1}^{n} p_{ij} = 1, \; (i = 1, \cdots, n),$$

(see **II.5.13**).

3.4.1 If $P = (p_{ij})$ is a row stochastic matrix, λ_t is a characteristic root of P and $\omega = \min_i \,(p_{ii})$, then

$$|\lambda_t - \omega| \leq 1 - \omega.$$

Proof: Consider $A = (a_{ij})$ where $a_{ij} = p_{ij} - \delta_{ij}$ in which δ_{ij} is the Kronecker delta (see **I.1.1**). Then A satisfies the conditions of **3.3.1** and thus, for any characteristic root μ_k of A,

$$|\mu_k + \epsilon| \leq \epsilon$$

where $\epsilon = \max_i (-a_{ii})$ as before. Now, $\lambda_t - 1$ is a characteristic root of A and therefore $|\lambda_t - (1 - \epsilon)| \leq 1 - (1 - \epsilon)$. But

$$1 - \epsilon = 1 - \max_i (-p_{ii} + 1) = \min_i p_{ii} = \omega.$$

Hence the result.

3.4.2   Let a_{ii} and a_{jj} be the smallest main diagonal elements of a row stochastic matrix $A = (a_{ij})$. Then all the characteristic roots of A lie in the interior or on the boundary of the oval

$$|z - a_{ii}| \, |z - a_{jj}| \leq (1 - a_{ii})(1 - a_{jj}).$$

3.4.3   Let λ_t be a characteristic root of a row stochastic matrix in $M_k(\mathbf{R})$, $(k \leq n)$. If $|arg(\lambda_t)| \leq 2\pi/n$, then λ_t lies in the interior or on the boundary of the regular polygon whose vertices are the nth roots of unity.

3.4.4   Let $A = (a_{ij}) \in M_n(\mathbf{R})$, be a stochastic matrix and let $m = \min_{i,j} (a_{ij})$. If

$$a_{ii} + a_{jj} > 1 - (n - 2)m, \qquad (i, j = 1, \cdots, n; i \neq j),$$

then each characteristic root of A has a positive real part, and the determinant of A is positive.

3.5   Normal matrices

Recall that a matrix $A = (a_{ij}) \in M_n(\mathbf{C})$ is normal if $AA^* = A^*A$ (see **I.4.7.1**). Alternatively, A is normal if and only if there exists a unitary matrix such that UAU^* is diagonal (see **I.4.10.3**).

We state, mostly without proof, some results on the localization of characteristic roots of normal matrices.

3.5.1   Every normal matrix $A = (a_{ij})$ has at least one characteristic root λ_t such that $|\lambda_t| \geq \max_{i,j} |a_{ij}|$ and $|\lambda_t| \geq \max |\tau_k|/k$ where τ_k is the sum of elements in any k-square principal submatrix of A.

3.5.2   If $A = (a_{ij}) \in M_n(\mathbf{C})$ is normal with characteristic roots $\lambda_t = \alpha_t + i\beta_t$, $(t = 1, \cdots, n; \alpha_t \in \mathbf{R}; \beta_t \in \mathbf{R})$, then

$$\max_t |\alpha_t| \geq \max_{i,j} |\tfrac{1}{2}(a_{ij} + \bar{a}_{ji})|, \qquad \max_t |\beta_t| \geq \max_{i,j} |\tfrac{1}{2}(a_{\cdot j} - \bar{a}_{\cdot i})|$$

and also

$$\max |\alpha_i| \geq \max |(\tau_k + \tau_k)/2k|, \qquad \max |\beta_i| \geq |(\tau_k - \tau_k)/2k|$$

where τ_k is defined as in **3.5.1**.

3.5.3 Let $A \in M_n(\mathbf{C})$ be normal and y be an n-vector such that $(Ay)_i = 0$ whenever $y_i = 0$. Then every circle containing all the n numbers $(Ay)_i/y_i$, $(y_i \neq 0)$, also contains a characteristic root of A.

3.5.4 Let $A \in M_n(\mathbf{C})$ be a normal matrix and y a real n-vector of unit length. Then in every closed half-plane, bounded by a straight line through the point (Ay, y), lies at least one characteristic root of A.

3.5.5 If A and y are defined as in **3.5.4** and δ is the nonnegative number defined by $\delta^2 = \|Ay\|^2 - |(Ay, y)|^2$, then each of the closed regions in which the circle $|z - (Ay, y)| = \delta$ divides the plane contains a characteristic root of A. This result will be proved in a more general form in **3.5.8**.

3.5.6 Let $A \in M_n(\mathbf{C})$ be normal and $x \in \mathbf{C}^n$ be of unit length. Let $\gamma \in \mathbf{C}$ and suppose ρ is the nonnegative real number defined by $\rho^2 = |\gamma - (Ax, x)|^2 + \|Ax\|^2 - |(Ax, x)|^2$. Then the circle with center γ and radius ρ contains a characteristic root of A.

3.5.7 Let $A \in M_n(\mathbf{C})$ be normal with characteristic roots $\lambda_1, \cdots, \lambda_n$. Let $x_1, \cdots, x_k$, $(1 \leq k \leq n)$, be orthonormal n-vectors, γ a complex number, and δ a nonnegative real number such that

$$\sum_{j=1}^{k} \|(A - \gamma I_n)x_j\|^2 \leq \delta. \tag{1}$$

Then

$$\sum_{i=1}^{k} |\lambda_{t_i} - \gamma|^2 \leq \delta \tag{2}$$

for some $(t_1, \cdots, t_k) \in Q_{k,n}$. The theorem remains true if both inequality signs in **3.5(1)** and **3.5(2)** are reversed.

Proof: The matrix $A - \gamma I_n$ is normal with characteristic roots $\lambda_i - \gamma$, $(i = 1, \cdots, n)$ (see **I.2.15.3**). Choose $t_1, \cdots, t_n$ so that

$$|\lambda_{t_1} - \gamma| \leq \cdots \leq |\lambda_{t_n} - \gamma|.$$

Now, $(A - \gamma I_n)(A - \gamma I_n)^*$ is hermitian with characteristic roots $|\lambda_{t_i} - \gamma|^2$, $(i = 1, \cdots, n)$. Hence

$$\sum_{i=1}^{k} |\lambda_{t_i} - \gamma|^2 = \min \sum_{j=1}^{k} \|(A - \gamma I_n)x_j\|^2 \tag{3}$$

where the minimum is taken over all orthonormal sets of k vectors in $\mathbf{C}^n$ (see **II.4.1.5**). Formula **3.5(3)** with **3.5(1)** implies **3.5(2)**. The last part of the theorem is proved similarly.

3.5.8 Let A, $\lambda_1, \cdots, \lambda_n$, $x_1, \cdots, x_k$ be defined as in **3.5.7**. Then there exists $(t_1, \cdots, t_k) \in Q_{k,n}$ such that

$$\sum_{i=1}^{k} \left|\lambda_{t_i} - \frac{1}{k}\sum_{j=1}^{k}(Ax_j, x_j)\right|^2 \leq \sum_{j=1}^{k}||Ax_j||^2 - \frac{1}{k}\left|\sum_{j=1}^{k}(Ax_j, x_j)\right|^2. \tag{4}$$

There also exists $(t_1, \cdots, t_k) \in Q_{k,n}$ satisfying the reversed inequality of **3.5(4)**.

Proof: In **3.5.7** set

$$\gamma = \frac{1}{k}\sum_{j=1}^{k}(Ax_j, x_j),$$

$$\delta = \sum_{j=1}^{k}||Ax_j||^2 - \frac{1}{k}\left|\sum_{j=1}^{k}(Ax_j, x_j)\right|^2.$$

It follows that $\sum_{j=1}^{k}||(A - \gamma I_n)x_j||^2 = \delta$ and **3.5(4)** follows from the first part of **3.5.7**. The second part of **3.5.7** implies the existence of subscripts satisfying the reversed inequality of **3.5(4)**.

When $k = 1$, the result in **3.5.8** reduces to **3.5.5**. For, **3.5(4)** reduces to $|\lambda_{t_1} - (Ax_1, x_1)| \leq ||Ax_1||^2 - |(Ax_1, x_1)|^2$.

3.5.9 Let $A \in M_m(\mathbf{C})$, $B \in M_n(\mathbf{C})$ be normal. Let $x \in \mathbf{C}^m$, $y \in \mathbf{C}^n$ be nonzero vectors such that $(Ax, x) = (By, y)$ and $||Ax|| \geq ||By||$. Then a circular disc in the complex plane containing all m characteristic roots of A contains at least one characteristic root of B.

3.5.10 If A and B are normal with characteristic roots $\alpha_1, \cdots, \alpha_n$ and $\beta_1, \cdots, \beta_n$ respectively, then there exists an ordering $\beta_{\sigma(1)}, \cdots, \beta_{\sigma(n)}$ such that $\sum_{i=1}^{n}|\alpha_i - \beta_{\sigma(i)}|^2 \leq ||A - B||^2$ where $||X||$ denotes the Euclidean norm of the matrix X (see **I.2.8.2**).

3.5.11 If A is a normal matrix with characteristic roots $\alpha_1, \cdots, \alpha_n$ and k is a positive real number, then $\beta_1, \cdots, \beta_n$ are characteristic roots of a normal matrix B satisfying $||A - B|| \leq k$ if and only if

$$\min_{\sigma \in S_n}\left(\sum_{i=1}^{n}|\alpha_i - \beta_{\sigma(i)}|^2\right)^{1/2} \leq k.$$

3.5.12 Let $\alpha_1, \cdots, \alpha_n$, $\beta_1, \cdots, \beta_n$ be arbitrary complex numbers and let Λ be the set of all numbers which can occur as characteristic roots of $A + B$, where A and B run over all normal n-square matrices with characteristic roots $\alpha_1, \cdots, \alpha_n$ and $\beta_1, \cdots, \beta_n$ respectively. Then Λ can be represented as an intersection:

$$\Lambda = \bigcap_{\Gamma} \{\alpha_i + \gamma \,|\, (i = 1, \cdots, n; \gamma \in \Gamma)\}$$

where Γ runs over all "circular" regions of the complex plane,

$$\{x + iy | ax + by + c(x^2 + y^2) + d \geq 0, \ (a, b, c, d, x, y \in \mathbf{R})\},$$

which contain $\beta_1, \cdots, \beta_n$.

3.5.13 In the special case of **3.5.9** when $\alpha_1, \cdots, \alpha_n$ are real and $\beta_1, \cdots, \beta_n$ are pure imaginary, it follows that

$$\Lambda = \bigcap \Delta$$

where Δ runs over all "hyperbolic" regions of the complex plane,

$$\{x + iy | ax + by + c(x^2 - y^2) + d \geq 0, \ (a, b, c, d, x, y \in \mathbf{R})\},$$

which contain the numbers $\alpha_i + \beta_j$, $(i, j = 1, \cdots, n)$.

3.6   Hermitian matrices

All the characteristic roots of a hermitian matrix are real (see **I.4.7.18**). Moreover, a hermitian matrix is normal and therefore the results of **3.5** apply to it. We specialize only one of these, **3.5.3**, because of the historical interest; the version of **3.5.3** specialized to hermitian matrices was the first result of this kind. The rest of this section consists of more recent results. We quote them without proof.

3.6.1   If A is a symmetric real matrix and $u = (u_1, \cdots, u_n)$ an n-tuple of nonzero real numbers, then in any interval containing the numbers $(Au)_i/u_i$, $(i = 1, \cdots, n)$, lies at least one characteristic root of A (see **3.5.3**).

3.6.2   Let $A = (a_{ij}) \in M_n(\mathbf{C})$ be hermitian with characteristic roots $\lambda_1 \geq \cdots \geq \lambda_n$. Let $c_1, \cdots, c_{n-1}$ and $d_1, \cdots, d_n$ be $2n - 1$ real numbers such that

$$c_i > 1, \qquad d_i \geq \frac{c_i^2}{c_i^2 - 1} \, d_{i+1}, \qquad (i = 1, \cdots, n - 1), \tag{1}$$

$$d_i \geq |a_{ii}| + c_i \Big(\sum_{j > i} |a_{ij}|^2 \Big)^{1/2}, \qquad (i = 1, \cdots, n). \tag{2}$$

Then

$$\lambda_i \leq d_i, \qquad (i = 1, \cdots, n). \tag{3}$$

3.6.3   The theorem **3.6.2** holds, if conditions **3.6(1)** and **3.6(2)** are replaced by

$$c_i > 0, \qquad d_i - d_{i+1} \geq \frac{1}{c_i}, \qquad (i = 1, \cdots, n - 1),$$

$$d_i \geq a_{ii} + c_i \sum_{j > i} |a_{ij}|^2, \qquad (i = 1, \cdots, n).$$

3.6.4   The hypothesis of **3.6.2** implies also that

$$\lambda_i \geq -d_{n-i+1}, \qquad (i = 1, \cdots, n).$$

3.6.5 Let $A = (a_{ij}) \in M_n(\mathbf{C})$, $(n \geq 2)$, be a hermitian matrix with characteristic roots $\lambda_1 \geq \cdots \geq \lambda_n$. Let $p_1, \cdots, p_s$ be any s integers, $(1 \leq s \leq n + 1)$, such that $0 = p_0 < p_1 < \cdots < p_s < p_{s+1} = n$. Define a hermitian matrix $B = (b_{ij}) \in M_n(\mathbf{C})$:

$$b_{ij} = \begin{cases} a_{ij}, & \text{if for some } k, \ 1 \leq k \leq s+1, \ p_{k-1} < i, j \leq p_k; \\ 0, & \text{otherwise.} \end{cases}$$

Thus B is a direct sum of nonoverlapping principal submatrices of A. Let $\mu_1 \geq \cdots \geq \mu_n$ be the characteristic roots of B. Then

$$\sum_{i=1}^{h} \mu_i \leq \sum_{i=1}^{h} \lambda_i, \qquad (h = 1, \cdots, n). \tag{3}$$

If, in addition, A is positive definite, then

$$\prod_{i=1}^{h} \lambda_{n-i+1} \leq \prod_{i=1}^{h} \mu_{n-i+1}, \qquad (h = 1, \cdots, n). \tag{4}$$

The Hadamard determinant theorem (see **II.4.1.7**) is a special case of 3.6(4).

3.6.6 Let $A; B; s; p_0, \cdots, p_{s+1}; \lambda_1, \cdots, \lambda_n;$ and $\mu_1, \cdots, \mu_n$ be defined as in **3.6.5**. For each $0 < i \leq p_s$ define b_i as follows:

$$b_i = \left(\sum_{j > p_k} |a_{ij}|^2 \right)^{1/2}$$

where $p_{k-1} < i \leq p_k$. Let $\beta_1 \geq \cdots \geq \beta_{p_s}$ be the numbers $b_1, \cdots, b_{p_s}$ rearranged in descending order and let $\gamma_i = \beta_i + \cdots + \beta_{p_s}$, $(i = 1, \cdots, p_s)$. Then

(i) $\lambda_i \leq \mu_j + \gamma_{i-j+1}$, if $1 \leq i - j + 1 \leq p_s$;

(ii) $\lambda_i \leq \mu_{i-p_s}$, if $i - p_s \geq 1$;

(iii) $\sum\limits_{i=1}^{h} (\lambda_i - \mu_i) \leq \gamma_1$, $(h = 1, \cdots, n)$;

(iv) $\lambda_i \geq \mu_j - \gamma_{j-i+1}$, if $1 \leq j - i + 1 \leq p_s$;

(v) $\lambda_i \geq \mu_{i+p_s}$, if $i + p_s \leq n$.

3.6.7 Let $\lambda_1 \geq \cdots \geq \lambda_n$ and $\mu_1 \geq \cdots \geq \mu_n$ and let N and K be the sets consisting of all n-tuples

$$(\lambda_1 + \mu_{\varphi(1)}, \cdots, \lambda_n + \mu_{\varphi(n)}) \quad \text{and} \quad (\lambda_{\varphi(1)} + \mu_1, \cdots, \lambda_{\varphi(n)} + \mu_n)$$

respectively as φ runs over all permutations $\varphi \in S_n$. Let $H(N)$ and $H(K)$ be the convex hulls of N and K (see **II.1.5**) and let $L = H(N) \cap H(K)$. If A and B are any symmetric real matrices with characteristic roots $\lambda_1, \cdots, \lambda_n$ and $\mu_1, \cdots, \mu_n$ respectively and if $\sigma_1 \geq \cdots \geq \sigma_n$ are the characteristic roots of $A + B$, then $(\sigma_1, \cdots, \sigma_n) \in L$.

If, in addition, either

$$|\mu_k - \mu_l| < |\lambda_i - \lambda_j|, \qquad (i, j, k, l = 1, \cdots, n; \ i \neq j),$$

$$\text{or} \qquad |\lambda_i - \lambda_j| < |\mu_k - \mu_l|, \qquad (i, j, k, l = 1, \cdots, n; \ k \neq l),$$

then

(i) L coincides with the set of all $(\sigma_1, \cdots, \sigma_n)$ as A and B vary over all symmetric real matrices with characteristic roots $\lambda_1, \cdots, \lambda_n$ and $\mu_1, \cdots, \mu_n$ respectively,

(ii) $\sigma_i \neq \sigma_j$ for $i \neq j$.

3.7 Jacobi, or triple diagonal, matrices

A matrix $A = (a_{ij}) \in M_n(\mathbf{C})$ is a *Jacobi matrix* (or a *triple diagonal matrix*), if $a_{ij} = 0$ whenever $|i - j| \geq 2$, (see **I.5.2**). Let

$$L_n = \begin{pmatrix} b_1 & c_1 & 0 & 0 & \cdots\cdots\cdots & 0 \\ a_2 & b_2 & c_2 & 0 & \cdots\cdots\cdots & 0 \\ 0 & a_3 & b_3 & c_3 & 0\cdots\cdots\cdots & 0 \\ \cdots\cdots\cdots\cdots\cdots\cdots\cdots\cdots\cdots \\ 0 & \cdots\cdots\cdots0 & a_{n-1} & b_{n-1} & c_{n-1} \\ 0 & \cdots\cdots\cdots0 & 0 & a_n & b_n \end{pmatrix}$$

be an n-square complex Jacobi matrix and let L_r denote the principal submatrix $L_n[1, \cdots, r | 1, \cdots, r]$.

3.7.1 If $L_n \in M_n(\mathbf{R})$ and $a_i c_{i-1} > 0$, $(i = 2, \cdots, n)$, then

(i) all characteristic roots of L_n are real and simple;

(ii) between any two characteristic roots of L_n lies exactly one characteristic root of L_{n-1};

(iii) if $\lambda_1 > \cdots > \lambda_n$ are the characteristic roots of L_n, then

$$d(\lambda_t I_{n-1} - L_{n-1})$$

is alternately positive and negative, $(t = 1, \cdots, n)$.

3.7.2 If $L_n \in M_n(\mathbf{R})$ and $a_i c_{i-1} < 0$, $(i = 2, \cdots, n)$, then all the real characteristic roots of L_n lie between the least and the greatest of the b_i, these values included. If, in addition, $b_1 < \cdots < b_n$, then the characteristic roots of L_n and those of L_{n-1} cannot interlace; in fact between any two adjacent real characteristic roots of L_n must lie an even number of real characteristic roots of L_{n-1}, if any.

3.7.3 If $L_n \in M_n(\mathbf{C})$, $b_2, \cdots, b_{n-1}$ are purely imaginary, $Re(b_1) \neq 0$, $a_j = -1$, $(j = 2, \cdots, n)$, and c_j, $(j = 1, \cdots, n-1)$, are nonzero real numbers, then the number of characteristic roots of L_n with positive real parts is equal to the number of positive elements in the sequence $c_0, c_0 c_1, c_0 c_1 c_2, \cdots, c_0 c_1 \cdots c_{n-1}$ where $c_0 = Re(b_1)$.

3.7.4 If $L_n \in M_n(\mathbf{C})$, $b_1, \cdots, b_{n-1}$ are purely imaginary, $Re(b_n) \neq 0$, $a_j = -1$, $(j = 2, \cdots, n)$, and c_j, $(j = 1, \cdots, n-1)$, are nonzero real numbers, then the number of characteristic roots of L_n with positive real parts is equal to the number of positive elements in the sequence c_n, $c_n c_{n-1}$, $\cdots$, $c_n c_{n-1} \cdots c_1$ where $c_n = Re(b_n)$. In particular, if $c_1, c_2, \cdots, c_{n-1}$, $-c_n$ are positive, then B is stable (see **3.3**).

4   The Spread of a Matrix

4.1   Definition

Let $A \in M_n(\mathbf{C})$, $(n \geq 3)$, and let $\lambda_1, \cdots, \lambda_n$ be the characteristic roots of A. The *spread* of A, written $s(A)$, is defined by: $s(A) = \max\limits_{i,j} |\lambda_i - \lambda_j|$.

4.2   Spread of a general matrix

4.2.1   If $A \in M_n(\mathbf{C})$ and $s(A)$ is defined as in **4.1**, then

$$s(A) \leq (2||A||^2 - \frac{2}{n} |\text{tr } A|^2)^{1/2}$$

with equality, if and only if A is a normal matrix with $n - 2$ of its characteristic roots equal to the arithmetic mean of the remaining two.

4.2.2 $s(A) \leq \sqrt{2}||A||$.

4.2.3   If $A \in M_n(\mathbf{R})$ and $s(A)$ is defined as in **4.1**, then

$$s(A) \leq \left(2\left(1 - \frac{1}{n}\right)(\text{tr } A)^2 - 4E_2(A)\right)^{1/2}$$

with equality, if and only if $n - 2$ of the characteristic roots of A are equal to the arithmetic mean of the remaining two.

4.3   Spread of a normal matrix

4.3.1   If $A \in M_n(\mathbf{C})$ is normal and $s(A)$ is defined as in **4.1**, then

$$s(A) \geq \sqrt{3} \max\limits_{i \neq j} |a_{ij}|.$$

4.3.2   If $A \in M_n(\mathbf{C})$ is hermitian and $s(A)$ is defined as in **4.1**, then

$$s(A) \geq 2 \max\limits_{i \neq j} |a_{ij}|.$$

This inequality is the best possible, in the sense that there exist hermitian matrices whose spread is equal to the absolute value of an off-diagonal element; e.g., if $A = \begin{pmatrix} 1 & -1 \\ -1 & 1 \end{pmatrix} \dotplus I_{n-2}$, then $s(A) = 2 = 2|a_{12}|$.

4.3.3 If $A \in M_n(\mathbf{C})$ is normal, then

(i) $s(A) \geq \max_{r \neq s} ((Re(a_{rr}) - Re(a_{ss}))^2 + |a_{rs} + \bar{a}_{sr}|^2)^{1/2},$

(ii) $s(A) \geq \max_{r \neq s} (|a_{rr} - a_{ss}|^2 + (|a_{rs}| - |a_{sr}|)^2)^{1/2},$

(iii) $s(A) \geq \max_{r \neq s} (|a_{rs}| + |a_{sr}|),$

(iv) $s(A) \geq \max_{r \neq s} (\frac{1}{2}(|a_{rr} - a_{ss}|^2 + |(a_{rr} - a_{ss})^2 + 4a_{rs}a_{sr}|$
$$+ 2|a_{rs}|^2 + 2|a_{sr}|^2))^{1/2}.$$

4.3.4 If $A \in M_n(\mathbf{C})$ is hermitian, then

$$s(A) \geq \max_{r \neq s} ((a_{rr} - a_{ss})^2 + 4|a_{rs}|^2)^{1/2}.$$

5   The Field of Values of a Matrix

5.1   Definitions

Let $A \in M_n(\mathbf{C})$. The *field of values* of A is just the totality of complex numbers $z = (Ax, x)$, $x \in \mathbf{C}^n$, $||x|| = 1$. The inner product in $\mathbf{C}^n$ is the one in **I.4.4**:

$$(x, y) = \sum_{i=1}^{n} x_i \bar{y}_i.$$

Thus the field of values is the image of the unit sphere under the mapping that associates with each unit vector x the complex number (Ax, x).

5.2   Properties of the field of values

5.2.1 For any matrix $A \in M_n(\mathbf{C})$ the boundary of the field of values is a convex curve and the field of values fills the interior of this curve. Thus the field of values is a convex set in the plane.

5.2.2 If A is a normal matrix, then the field of values of A is just the convex hull of the set of characteristic roots of A, $H(\lambda(A))$.

5.2.3 If $A \in M_n(\mathbf{C})$ and $H(\lambda(A))$ is the field of values of A and $n \leq 4$, then A is normal. If $n \geq 5$, there exist matrices $A \in M_n(\mathbf{C})$ for which $H(\lambda(A))$ is the field of values of A but A is not normal.

5.2.4 If $A \in M_n(\mathbf{C})$, then $H(\lambda(A))$ is always a subset of the field of values of A. For, if $r \in \lambda(A)$ and u is a corresponding characteristic vector of length 1, then $r = (Au, u)$. Since the field of values is convex the result follows.

5.2.5 If U is a unitary matrix in $M_n(\mathbf{C})$ and $A \in M_n(\mathbf{C})$, then the field of values of A and U^*AU are the same. For $(U^*AUx, x) = (AUx, Ux)$

and $(Ux, Ux) = (U^*Ux, x) = (x, x)$. Moreover, as x ranges over all unit vectors so does Ux.

5.2.6 If $A \in M_n(\mathbf{C})$, then A is hermitian if and only if the field of values of A is an interval on the real line.

The necessity is an immediate consequence of **5.2.2**. To see the sufficiency let $A = H + iK$ in which H and K are both hermitian (see **I.4.17.15**). Then $(Ax, x) = (Hx, x) + i(Kx, x)$ and $(Kx, x) = 0$ for every vector x. It follows that $K = 0_{n,n}$ and $A = H$ is hermitian.

5.2.7 If $A \in M_n(\mathbf{C})$, then the field of values of A is a subset of the rectangle in the complex plane whose four vertices are (α_n, β_n), (α_n, β_1), (α_1, β_n), (α_1, β_1): as in **5.2.6**, $A = H + iK$, α_1 and β_1 are the largest characteristic roots of H and K respectively, whereas α_n and β_n are the smallest characteristic roots of H and K respectively.

5.2.8 If r is a characteristic root of $A \in M_n(\mathbf{C})$ and $(\lambda - r)^p$, $(p > 1)$, is an elementary divisor of $\lambda I_n - A$, then r is strictly in the interior of the field of values of A.

References

§§1.2.1, 1.2.2. [B2].
§§1.3.1, 1.3.2. [B19].
§1.3.3. [B5].
§§1.4.1, 1.4.2. [B42].
§1.5.1. [B6].
§1.6.1. [B38].
§1.6.2. [B16].
§1.6.3. [B7].
§1.6.4. [B36]; [B14].
§1.6.5. [B3], I.
§1.6.6. [A7], p. 36; [B3], I.
§1.7. [B40].
§2.1.1. [B3], I; [B17]; [B23]; [B9]; [A3], pp. 13–14; [B26]; [B47].
§2.1.2. [B23]; [B9].
§2.2.1. [B17].
§2.2.2. [B3], I; also see §1.6.6.
§2.2.3. [B3], I.
§2.2.4. [B47].
§2.2.5. [B17].
§2.2.6. [B46].

§2.2.7. [B33].

§2.2.8. [B44].

§§2.2.9, 2.2.10. [B41].

§§2.2.11–2.2.13. [B13].

§2.4.1. This result is due to Ostrowski (see [B30]). The method of proof follows Brauer's proof of **2.4.2** (see [B3], II, Theorem 11).

§2.4.2. [B3], II.

§§2.5.1–2.5.7. [B31].

§2.6. [B21]; [B12].

§3.1.1. [B16].

§3.1.2. [B22].

§3.1.3. [B32].

§3.1.4. [B4].

§3.1.5. [B34].

§3.1.6. [B25].

§§3.3.1, 3.3.2. [B45].

§3.3.3. [B39].

§3.3.4. [B15].

§3.3.6. [B40].

§3.3.7. [A4], pp. 276–277; [B35].

§3.3.8. [B35].

§3.3.9. [B48].

§3.4.1. [B15].

§3.4.2. [B3], IV.

§3.4.3. [B10].

§3.4.4. [B3], IV.

§§3.5.1, 3.5.2. [B37].

§3.5.3. [B50]; [B51].

§§3.5.4, 3.5.5. [B51].

§3.5.6. [B50].

§§3.5.7–3.5.9. [B13].

§§3.5.10, 3.5.11. [B20].

§§3.5.12, 3.5.13. [B52].

§3.6.1. [B8].

§§3.6.2–3.6.6. [B11].

§3.6.7. [B24].

§§3.7.1, 3.7.2. [B1].

§§3.7.3, 3.7.4. [B43].

§§4.2.1–4.2.3. [B27].

§4.3.1. *Ibid.*

§4.3.2. [B37].

§§4.3.3, 4.3.4. [B28].

§§5.2.1, 5.2.2. [B49].

§5.2.3. [B29].

§5.2.4. [B49].

§5.2.7. *Ibid.*

§5.2.8. [B18].

Bibliography

Part A. Books

1. Bellman, R., *Introduction to matrix analysis*, McGraw-Hill, New York (1960).
2. Gantmacher, F. R., *The theory of matrices*, vols. I, II (trans., K. A. Hirsch), Chelsea, New York (1959).
3. Hadamard, J., *Leçons sur la propagation des ondes*, Chelsea, New York (1949).
4. Lyapunov, A., *Problème général de la stabilité du mouvement*, Ann. of Math. Studies *17*, Princeton University, Princeton, N.J. (1947).
5. MacDuffee, C. C., *The theory of matrices*, Chelsea, New York (1946).
6. Parodi, M., *La localisation des valeurs caractéristiques des matrices et ses applications*, Gauthier-Villars, Paris (1959).
7. Perron, O., *Theorie der algebraischen Gleichungen*, II (zweite Auflage), de Gruyter, Berlin (1933).
8. Todd, J. (ed.), *Survey of numerical analysis*, McGraw-Hill, New York (1962).

Part B. Papers

1. Arscott, F. M., *Latent roots of tri-diagonal matrices*, Proc. Edinburgh Math. Soc., *12* (1961), Edinburgh Math. Notes No. 44, 5–7.
2. Bendixson, I., *Sur les racines d'une équation fondamentale*, Acta Math., *25* (1902), 359–365. (Öfversigt af K. Vet. Akad. Forh. Stockholm, *57* (1900), 1099.)
3. Brauer, A., *Limits for the characteristic roots of a matrix*, Duke Math. J., I, *13* (1946), 387–395; II, *14* (1947), 21–26; IV, *19* (1952), 73–91.
4. Brauer, A., *The theorems of Ledermann and Ostrowski on positive matrices*, Duke Math. J., *24* (1957); 265–274.
5. Bromwich, T. J. I'A, *On the roots of the characteristic equation of a linear substitution*, Acta. Math., *30* (1906), 297–304.
6. Browne, E. T., *The characteristic equation of a matrix*, Bull. Amer. Math. Soc., *34* (1928), 363–368.
7. Browne, E. T., *The characteristic roots of a matrix*, Bull. Amer. Math. Soc., *36* (1930), 705–710.
8. Collatz, L., *Einschliessungssatz für die charakteristischen Zahlen von Matrizen*, Math. Z., *48* (1942), 221–226.
9. Desplanques, J., *Théorème d'algèbre*, J. de Math. Spec., *9* (1887), 12–13.
10. Dmitriev, N. A., and Dynkin, E. B., *On the characteristic roots of stochastic matrices*, Dokl. Akad. Nauk SSSR, *49* (1945), 159–162.
11. Fan, Ky, *Inequalities for eigenvalues of Hermitian matrices*, Nat. Bur. Standards, Appl. Math. Ser., *39* (1954), 131–139.

12. Fan, Ky, *Note on circular disks containing the eigenvalues of a matrix*, Duke Math. J., *25* (1958), 441–445.

13. Fan, Ky, and Hoffman, A. J., *Lower bounds for the rank and location of the eigenvalues of a matrix*, Nat. Bur. Standards. Appl. Math. Ser., *39* (1954), 117–130.

14. Farnell, A. B., *The characteristic roots of a matrix*, Bull. Amer. Math. Soc., *50* (1944), 789–794.

15. Fréchet, M., *Comportment asymptotique des solutions d'un système d'équations linéaires et homogènes aux différences finies du premier ordre à coefficients constants*, Publ. Fac. Sci. Univ. Masaryk; No. 178 (1933), 1–24.

16. Frobenius, G., *Über Matrizen aus positiven Elementen*, S.-B. Kgl. Preuss. Akad. Wiss., (1908), 471–476.

17. Geršgorin, S. A., *Über die Abgrenzung der Eigenwerte einer Matrix*, Izv. Akad. Nauk SSSR, Ser. Fiz.-Mat., *6* (1931), 749–754.

18. Givens, W., *Fields of values of a matrix*, Proc. Amer. Math. Soc., *3* (1952), 206–209.

19. Hirsch, A., *Sur les racines d'une équation fondamentale*, Acta Math., *25* (1902), 367–370.

20. Hoffman, A. J., and Wielandt, H. W., *The variation of the spectrum of a normal matrix*, Duke Math. J., *20* (1953), 37–39.

21. Kotelyanskiĭ, D. M., *On the disposition of the points of a matrix spectrum*, Ukrain. Math. Ž., *7* (1955), 131–133 (Russian).

22. Ledermann, W., *Bounds for the greatest latent root of a positive matrix*, J. London Math. Soc., *25* (1950), 265–268.

23. Lévy, L., *Sur la possibilité du l'équilibre électrique*, C. R. Acad. Sci. Paris, *93* (1881), 706–708.

24. Lidskiĭ, V. C., *The proper values of the sum and products of symmetric matrices*, Dokl. Akad. Nauk SSSR, *75* (1950), 769–772. [Nat. Bur. Standards Report 2248, (1953).]

25. Marcus, M., Minc, H., and Moyls, B., *Some results on non-negative matrices*, J. Res. Nat. Bur. Standards, *65B* (1961), 205–209.

26. Minkowski, H., *Zur Theorie der Einheiten in den algebraischen Zahlkörpern*, Nachr. Kgl. Wiss. Göttingen, Math.-Phys. Kl., (1900), 90–93. (Ges. Abh., Leipzig (1911), I, 316–319.)

27. Mirsky, L., *The spread of a matrix*, Mathematika, *3* (1956), 127–130.

28. Mirsky, L., *Inequalities for normal and hermitian matrices*, Duke Math. J., *24* (1957), 591–599.

29. Moyls, B. N., and Marcus, M., *Field convexity of a square matrix*, Proc. Amer. Math. Soc., *6* (1955), 981–983.

30. Ostrowski, A., *Über die Determinanten mit überwiegender Hauptdiagonale*, Comm. Math. Helv., *10* (1937), 69–96.

31. Ostrowski, A., *Über das Nichtverschwinden einer Klasse von Determinanten und die Lokalisierung der charakteristischen Wurzeln von Matrizen*, Compositio Math., *9* (1951), 209–226.

32. Ostrowski, A., *Bounds for the greatest latent root of a positive matrix*, J. London Math. Soc., *27* (1952), 253–256.

33. Ostrowski, A., *On nearly triangular matrices*, J. Res. Nat. Bur. Standards, *52* (1954), 319–345.

34. Ostrowski, A., and Schneider, H., *Bounds for the maximal characteristic root of a non-negative irreducible matrix*, Duke Math. J., *27* (1960), 547–553.

35. Ostrowski, A., and Schneider, H., *Some theorems on the inertia of general matrices*, J. Math. Anal. and Appl., *4* (1962), 72–84.

36. Parker, W. V., *The characteristic roots of a matrix*, Duke Math. J., *3* (1937), 484–487.

37. Parker, W. V., *Numbers associated with a matrix*, Duke Math. J., *15* (1948), 711–715.

38. Pick, G., *Über die Wurzeln der charakteristischen Gleichungen von Schwingungs-problemen*, Z. Angew. Math. Mech., *2* (1922), 353–357.

39. Rohrbach, H., *Bemerkungen zu einem Determinantensatz von Minkowski*, Jber. Deutsch. Math. Verein., *40* (1931), 49–53.

40. Schneider, H., *An inequality for latent roots applied to determinants with dominant principal diagonal*, J. London Math. Soc., *28* (1953), 8–20.

41. Schneider, H., *Regions of exclusion for the latent roots of a matrix*, Proc. Amer. Math. Soc., *5* (1954), 320–322.

42. Schur, I., *Über die charakteristischen Wurzeln einer linearen Substitution mit einer Anwendung auf die Theorie der Integralgleichungen*, Math. Ann., *65* (1909), 488–510.

43. Schwarz, H.-R., *Ein Verfahren zur Stabilitätsfrage bei Matrizen-Eigenwert-Problemen*, Z. Angew. Math. Phys., *7* (1956), 473–500.

44. Stein, P., *A note on bounds of multiple characteristic roots of a matrix*, J. Res. Nat. Bur. Standards, *48* (1952), 59–60.

45. Tambs-Lyche, R., *Un théorème sur les déterminants*, Det. Kong. Vid. Selskab., Forh. I, nr. 41 (1928), 119–120.

46. Taussky, Olga, *Bounds for characteristic roots of matrices*, Duke Math. J., *15* (1948), 1043–1044.

47. Taussky, Olga, *A recurring theorem on determinants*, Amer Math. Monthly, *56* (1949), 672–676.

48. Taussky, Olga, *A generalization of a theorem of Lyapunov*, J. Soc. Indust. Appl. Math., *9* (1961), 640–643.

49. Toeplitz, O., *Das algebraische Analogon zu einem Satze von Féjer*, Math. Z., *2* (1918), 187–197.

50. Walker, A. G., and Weston, J. D., *Inclusion theorems for the eigenvalues of a normal matrix*, J. London Math. Soc., *24* (1949), 28–31.

51. Wielandt, H., *Ein Einschliessungssatz für charakteristische Wurzeln normaler Matrizen*, Arch. Math., *1* (1948), 348–352.

52. Wielandt, H., *On eigenvalues of sums of normal matrices*, Pacific J. Math., *5* (1955), 633–638.

Index

175

A CATALOG OF SELECTED
DOVER BOOKS
IN SCIENCE AND MATHEMATICS

A CATALOG OF SELECTED
DOVER BOOKS
IN SCIENCE AND MATHEMATICS

QUALITATIVE THEORY OF DIFFERENTIAL EQUATIONS, V.V. Nemytskii and V.V. Stepanov. Classic graduate-level text by two prominent Soviet mathematicians covers classical differential equations as well as topological dynamics and ergodic theory. Bibliographies. 523pp. 5⅜ × 8½. 65954-2 Pa. $10.95

MATRICES AND LINEAR ALGEBRA, Hans Schneider and George Phillip Barker. Basic textbook covers theory of matrices and its applications to systems of linear equations and related topics such as determinants, eigenvalues and differential equations. Numerous exercises. 432pp. 5⅜ × 8½. 66014-1 Pa. $9.95

QUANTUM THEORY, David Bohm. This advanced undergraduate-level text presents the quantum theory in terms of qualitative and imaginative concepts, followed by specific applications worked out in mathematical detail. Preface. Index. 655pp. 5⅜ × 8½. 65969-0 Pa. $13.95

ATOMIC PHYSICS (8th edition), Max Born. Nobel laureate's lucid treatment of kinetic theory of gases, elementary particles, nuclear atom, wave-corpuscles, atomic structure and spectral lines, much more. Over 40 appendices, bibliography. 495pp. 5⅜ × 8½. 65984-4 Pa. $12.95

ELECTRONIC STRUCTURE AND THE PROPERTIES OF SOLIDS: The Physics of the Chemical Bond, Walter A. Harrison. Innovative text offers basic understanding of the electronic structure of covalent and ionic solids, simple metals, transition metals and their compounds. Problems. 1980 edition. 582pp. 6⅛ × 9¼. 66021-4 Pa. $15.95

BOUNDARY VALUE PROBLEMS OF HEAT CONDUCTION, M. Necati Özisik. Systematic, comprehensive treatment of modern mathematical methods of solving problems in heat conduction and diffusion. Numerous examples and problems. Selected references. Appendices. 505pp. 5⅜ × 8½. 65990-9 Pa. $11.95

A SHORT HISTORY OF CHEMISTRY (3rd edition), J.R. Partington. Classic exposition explores origins of chemistry, alchemy, early medical chemistry, nature of atmosphere, theory of valency, laws and structure of atomic theory, much more. 428pp. 5⅜ × 8½. (Available in U.S. only) 65977-1 Pa. $10.95

A HISTORY OF ASTRONOMY, A. Pannekoek. Well-balanced, carefully reasoned study covers such topics as Ptolemaic theory, work of Copernicus, Kepler, Newton, Eddington's work on stars, much more. Illustrated. References. 521pp. 5⅜ × 8½. 65994-1 Pa. $12.95

PRINCIPLES OF METEOROLOGICAL ANALYSIS, Walter J. Saucier. Highly respected, abundantly illustrated classic reviews atmospheric variables, hydrostatics, static stability, various analyses (scalar, cross-section, isobaric, isentropic, more). For intermediate meteorology students. 454pp. 6⅛ × 9¼. 65979-8 Pa. $14.95

RELATIVITY, THERMODYNAMICS AND COSMOLOGY, Richard C. Tolman. Landmark study extends thermodynamics to special, general relativity; also applications of relativistic mechanics, thermodynamics to cosmological models. 501pp. 5⅜ × 8½. 65383-8 Pa. $12.95

APPLIED ANALYSIS, Cornelius Lanczos. Classic work on analysis and design of finite processes for approximating solution of analytical problems. Algebraic equations, matrices, harmonic analysis, quadrature methods, much more. 559pp. 5⅜ × 8½. 65656-X Pa. $12.95

SPECIAL RELATIVITY FOR PHYSICISTS, G. Stephenson and C.W. Kilmister. Concise elegant account for nonspecialists. Lorentz transformation, optical and dynamical applications, more. Bibliography. 108pp. 5⅜ × 8½. 65519-9 Pa. $4.95

INTRODUCTION TO ANALYSIS, Maxwell Rosenlicht. Unusually clear, accessible coverage of set theory, real number system, metric spaces, continuous functions, Riemann integration, multiple integrals, more. Wide range of problems. Undergraduate level. Bibliography. 254pp. 5⅜ × 8½. 65038-3 Pa. $7.95

INTRODUCTION TO QUANTUM MECHANICS With Applications to Chemistry, Linus Pauling & E. Bright Wilson, Jr. Classic undergraduate text by Nobel Prize winner applies quantum mechanics to chemical and physical problems. Numerous tables and figures enhance the text. Chapter bibliographies. Appendices. Index. 468pp. 5⅜ × 8½. 64871-0 Pa. $11.95

ASYMPTOTIC EXPANSIONS OF INTEGRALS, Norman Bleistein & Richard A. Handelsman. Best introduction to important field with applications in a variety of scientific disciplines. New preface. Problems. Diagrams. Tables. Bibliography. Index. 448pp. 5⅜ × 8½. 65082-0 Pa. $12.95

MATHEMATICS APPLIED TO CONTINUUM MECHANICS, Lee A. Segel. Analyzes models of fluid flow and solid deformation. For upper-level math, science and engineering students. 608pp. 5⅜ × 8½. 65369-2 Pa. $13.95

ELEMENTS OF REAL ANALYSIS, David A. Sprecher. Classic text covers fundamental concepts, real number system, point sets, functions of a real variable, Fourier series, much more. Over 500 exercises. 352pp. 5⅜ × 8½. 65385-4 Pa. $10.95

PHYSICAL PRINCIPLES OF THE QUANTUM THEORY, Werner Heisenberg. Nobel Laureate discusses quantum theory, uncertainty, wave mechanics, work of Dirac, Schroedinger, Compton, Wilson, Einstein, etc. 184pp. 5⅜ × 8½. 60113-7 Pa. $5.95

INTRODUCTORY REAL ANALYSIS, A.N. Kolmogorov, S.V. Fomin. Translated by Richard A. Silverman. Self-contained, evenly paced introduction to real and functional analysis. Some 350 problems. 403pp. 5⅜ × 8½. 61226-0 Pa. $9.95

PROBLEMS AND SOLUTIONS IN QUANTUM CHEMISTRY AND PHYSICS, Charles S. Johnson, Jr. and Lee G. Pedersen. Unusually varied problems, detailed solutions in coverage of quantum mechanics, wave mechanics, angular momentum, molecular spectroscopy, scattering theory, more. 280 problems plus 139 supplementary exercises. 430pp. 6½ × 9¼. 65236-X Pa. $12.95

ASYMPTOTIC METHODS IN ANALYSIS, N.G. de Bruijn. An inexpensive, comprehensive guide to asymptotic methods—the pioneering work that teaches by explaining worked examples in detail. Index. 224pp. 5⅜ × 8½.   64221-6 Pa. $6.95

OPTICAL RESONANCE AND TWO-LEVEL ATOMS, L. Allen and J.H. Eberly. Clear, comprehensive introduction to basic principles behind all quantum optical resonance phenomena. 53 illustrations. Preface. Index. 256pp. 5⅜ × 8½.
65533-4 Pa. $7.95

COMPLEX VARIABLES, Francis J. Flanigan. Unusual approach, delaying complex algebra till harmonic functions have been analyzed from real variable viewpoint. Includes problems with answers. 364pp. 5⅜ × 8½.   61388-7 Pa. $8.95

ATOMIC SPECTRA AND ATOMIC STRUCTURE, Gerhard Herzberg. One of best introductions; especially for specialist in other fields. Treatment is physical rather than mathematical. 80 illustrations. 257pp. 5⅜ × 8½.   60115-3 Pa. $5.95

APPLIED COMPLEX VARIABLES, John W. Dettman. Step-by-step coverage of fundamentals of analytic function theory—plus lucid exposition of five important applications: Potential Theory; Ordinary Differential Equations; Fourier Transforms; Laplace Transforms; Asymptotic Expansions. 66 figures. Exercises at chapter ends. 512pp. 5⅜ × 8½.   64670-X Pa. $11.95

ULTRASONIC ABSORPTION: An Introduction to the Theory of Sound Absorption and Dispersion in Gases, Liquids and Solids, A.B. Bhatia. Standard reference in the field provides a clear, systematically organized introductory review of fundamental concepts for advanced graduate students, research workers. Numerous diagrams. Bibliography. 440pp. 5⅜ × 8½.   64917-2 Pa. $11.95

UNBOUNDED LINEAR OPERATORS: Theory and Applications, Seymour Goldberg. Classic presents systematic treatment of the theory of unbounded linear operators in normed linear spaces with applications to differential equations. Bibliography. 199pp. 5⅜ × 8½.   64830-3 Pa. $7.95

LIGHT SCATTERING BY SMALL PARTICLES, H.C. van de Hulst. Comprehensive treatment including full range of useful approximation methods for researchers in chemistry, meteorology and astronomy. 44 illustrations. 470pp. 5⅜ × 8½.   64228-3 Pa. $10.95

CONFORMAL MAPPING ON RIEMANN SURFACES, Harvey Cohn. Lucid, insightful book presents ideal coverage of subject. 334 exercises make book perfect for self-study. 55 figures. 352pp. 5⅜ × 8¼.   64025-6 Pa. $9.95

OPTICKS, Sir Isaac Newton. Newton's own experiments with spectroscopy, colors, lenses, reflection, refraction, etc., in language the layman can follow. Foreword by Albert Einstein. 532pp. 5⅜ × 8½.   60205-2 Pa. $9.95

GENERALIZED INTEGRAL TRANSFORMATIONS, A.H. Zemanian. Graduate-level study of recent generalizations of the Laplace, Mellin, Hankel, K. Weierstrass, convolution and other simple transformations. Bibliography. 320pp. 5⅜ × 8½.   65375-7 Pa. $8.95

THE ELECTROMAGNETIC FIELD, Albert Shadowitz. Comprehensive undergraduate text covers basics of electric and magnetic fields, builds up to electromagnetic theory. Also related topics, including relativity. Over 900 problems. 768pp. 5⅜ × 8¼.
65660-8 Pa. $18.95

FOURIER SERIES, Georgi P. Tolstov. Translated by Richard A. Silverman. A valuable addition to the literature on the subject, moving clearly from subject to subject and theorem to theorem. 107 problems, answers. 336pp. 5⅜ × 8½.
63317-9 Pa. $8.95

THEORY OF ELECTROMAGNETIC WAVE PROPAGATION, Charles Herach Papas. Graduate-level study discusses the Maxwell field equations, radiation from wire antennas, the Doppler effect and more. xiii + 244pp. 5⅜ × 8½.
65678-0 Pa. $6.95

DISTRIBUTION THEORY AND TRANSFORM ANALYSIS: An Introduction to Generalized Functions, with Applications, A.H. Zemanian. Provides basics of distribution theory, describes generalized Fourier and Laplace transformations. Numerous problems. 384pp. 5⅜ × 8½.
65479-6 Pa. $9.95

THE PHYSICS OF WAVES, William C. Elmore and Mark A. Heald. Unique overview of classical wave theory. Acoustics, optics, electromagnetic radiation, more. Ideal as classroom text or for self-study. Problems. 477pp. 5⅜ × 8½.
64926-1 Pa. $12.95

CALCULUS OF VARIATIONS WITH APPLICATIONS, George M. Ewing. Applications-oriented introduction to variational theory develops insight and promotes understanding of specialized books, research papers. Suitable for advanced undergraduate/graduate students as primary, supplementary text. 352pp. 5⅜ × 8½.
64856-7 Pa. $8.95

A TREATISE ON ELECTRICITY AND MAGNETISM, James Clerk Maxwell. Important foundation work of modern physics. Brings to final form Maxwell's theory of electromagnetism and rigorously derives his general equations of field theory. 1,084pp. 5⅜ × 8½.
60636-8, 60637-6 Pa., Two-vol. set $19.90

AN INTRODUCTION TO THE CALCULUS OF VARIATIONS, Charles Fox. Graduate-level text covers variations of an integral, isoperimetrical problems, least action, special relativity, approximations, more. References. 279pp. 5⅜ × 8½.
65499-0 Pa. $7.95

HYDRODYNAMIC AND HYDROMAGNETIC STABILITY, S. Chandrasekhar. Lucid examination of the Rayleigh-Benard problem; clear coverage of the theory of instabilities causing convection. 704pp. 5⅜ × 8¼.
64071-X Pa. $14.95

CALCULUS OF VARIATIONS, Robert Weinstock. Basic introduction covering isoperimetric problems, theory of elasticity, quantum mechanics, electrostatics, etc. Exercises throughout. 326pp. 5⅜ × 8½.
63069-2 Pa. $7.95

DYNAMICS OF FLUIDS IN POROUS MEDIA, Jacob Bear. For advanced students of ground water hydrology, soil mechanics and physics, drainage and irrigation engineering and more. 335 illustrations. Exercises, with answers. 784pp. 6⅛ × 9¼.
65675-6 Pa. $19.95

NUMERICAL METHODS FOR SCIENTISTS AND ENGINEERS, Richard Hamming. Classic text stresses frequency approach in coverage of algorithms, polynomial approximation, Fourier approximation, exponential approxima-tion, other topics. Revised and enlarged 2nd edition. 721pp. 5⅜ × 8½.
65241-6 Pa. $14.95

THEORETICAL SOLID STATE PHYSICS, Vol. I: Perfect Lattices in Equilib-rium; Vol. II: Non-Equilibrium and Disorder, William Jones and Norman H. March. Monumental reference work covers fundamental theory of equilibrium properties of perfect crystalline solids, non-equilibrium properties, defects and disordered systems. Appendices. Problems. Preface. Diagrams. Index. Bibliog-raphy. Total of 1,301pp. 5⅜ × 8½. Two volumes. Vol. I 65015-4 Pa. $14.95
Vol. II 65016-2 Pa. $14.95

OPTIMIZATION THEORY WITH APPLICATIONS, Donald A. Pierre. Broad-spectrum approach to important topic. Classical theory of minima and maxima, calculus of variations, simplex technique and linear programming, more. Many problems, examples. 640pp. 5⅜ × 8½. 65205-X Pa. $14.95

THE MODERN THEORY OF SOLIDS, Frederick Seitz. First inexpensive edition of classic work on theory of ionic crystals, free-electron theory of metals and semiconductors, molecular binding, much more. 736pp. 5⅜ × 8½.
65482-6 Pa. $15.95

ESSAYS ON THE THEORY OF NUMBERS, Richard Dedekind. Two classic essays by great German mathematician: on the theory of irrational numbers; and on transfinite numbers and properties of natural numbers. 115pp. 5⅜ × 8½.
21010-3 Pa. $4.95

THE FUNCTIONS OF MATHEMATICAL PHYSICS, Harry Hochstadt. Com-prehensive treatment of orthogonal polynomials, hypergeometric functions, Hill's equation, much more. Bibliography. Index. 322pp. 5⅜ × 8½. 65214-9 Pa. $9.95

NUMBER THEORY AND ITS HISTORY, Oystein Ore. Unusually clear, accessible introduction covers counting, properties of numbers, prime numbers, much more. Bibliography. 380pp. 5⅜ × 8½. 65620-9 Pa. $9.95

THE VARIATIONAL PRINCIPLES OF MECHANICS, Cornelius Lanczos. Graduate level coverage of calculus of variations, equations of motion, relativistic mechanics, more. First inexpensive paperbound edition of classic treatise. Index. Bibliography. 418pp. 5⅜ × 8½. 65067-7 Pa. $11.95

MATHEMATICAL TABLES AND FORMULAS, Robert D. Carmichael and Edwin R. Smith. Logarithms, sines, tangents, trig functions, powers, roots, reciprocals, exponential and hyperbolic functions, formulas and theorems. 269pp. 5⅜ × 8½. 60111-0 Pa. $6.95

THEORETICAL PHYSICS, Georg Joos, with Ira M. Freeman. Classic overview covers essential math, mechanics, electromagnetic theory, thermodynamics, quan-tum mechanics, nuclear physics, other topics. First paperback edition. xxiii + 885pp. 5⅜ × 8½. 65227-0 Pa. $19.95

CATALOG OF DOVER BOOKS

HANDBOOK OF MATHEMATICAL FUNCTIONS WITH FORMULAS, GRAPHS, AND MATHEMATICAL TABLES, edited by Milton Abramowitz and Irene A. Stegun. Vast compendium: 29 sets of tables, some to as high as 20 places. 1,046pp. 8 × 10½. 61272-4 Pa. $24.95

MATHEMATICAL METHODS IN PHYSICS AND ENGINEERING, John W. Dettman. Algebraically based approach to vectors, mapping, diffraction, other topics in applied math. Also generalized functions, analytic function theory, more. Exercises. 448pp. 5⅜ × 8¼. 65649-7 Pa. $9.95

A SURVEY OF NUMERICAL MATHEMATICS, David M. Young and Robert Todd Gregory. Broad self-contained coverage of computer-oriented numerical algorithms for solving various types of mathematical problems in linear algebra, ordinary and partial, differential equations, much more. Exercises. Total of 1,248pp. 5⅜ × 8½. Two volumes. Vol. I 65691-8 Pa. $14.95
Vol. II 65692-6 Pa. $14.95

TENSOR ANALYSIS FOR PHYSICISTS, J.A. Schouten. Concise exposition of the mathematical basis of tensor analysis, integrated with well-chosen physical examples of the theory. Exercises. Index. Bibliography. 289pp. 5⅜ × 8½. 65582-2 Pa. $8.95

INTRODUCTION TO NUMERICAL ANALYSIS (2nd Edition), F.B. Hildebrand. Classic, fundamental treatment covers computation, approximation, interpolation, numerical differentiation and integration, other topics. 150 new problems. 669pp. 5⅜ × 8½. 65363-3 Pa. $14.95

INVESTIGATIONS ON THE THEORY OF THE BROWNIAN MOVEMENT, Albert Einstein. Five papers (1905–8) investigating dynamics of Brownian motion and evolving elementary theory. Notes by R. Fürth. 122pp. 5⅜ × 8½. 60304-0 Pa. $4.95

CATASTROPHE THEORY FOR SCIENTISTS AND ENGINEERS, Robert Gilmore. Advanced-level treatment describes mathematics of theory grounded in the work of Poincaré, R. Thom, other mathematicians. Also important applications to problems in mathematics, physics, chemistry and engineering. 1981 edition. References. 28 tables. 397 black-and-white illustrations. xvii + 666pp. 6⅛ × 9¼. 67539-4 Pa. $16.95

AN INTRODUCTION TO STATISTICAL THERMODYNAMICS, Terrell L. Hill. Excellent basic text offers wide-ranging coverage of quantum statistical mechanics, systems of interacting molecules, quantum statistics, more. 523pp. 5⅜ × 8½. 65242-4 Pa. $12.95

ELEMENTARY DIFFERENTIAL EQUATIONS, William Ted Martin and Eric Reissner. Exceptionally clear, comprehensive introduction at undergraduate level. Nature and origin of differential equations, differential equations of first, second and higher orders. Picard's Theorem, much more. Problems with solutions. 331pp. 5⅜ × 8½. 65024-3 Pa. $8.95

STATISTICAL PHYSICS, Gregory H. Wannier. Classic text combines thermodynamics, statistical mechanics and kinetic theory in one unified presentation of thermal physics. Problems with solutions. Bibliography. 532pp. 5⅜ × 8½. 65401-X Pa. $11.95

ORDINARY DIFFERENTIAL EQUATIONS, Morris Tenenbaum and Harry Pollard. Exhaustive survey of ordinary differential equations for undergraduates in mathematics, engineering, science. Thorough analysis of theorems. Diagrams. Bibliography. Index. 818pp. 5⅜ × 8½. 64940-7 Pa. $16.95

STATISTICAL MECHANICS: Principles and Applications, Terrell L. Hill. Standard text covers fundamentals of statistical mechanics, applications to fluctuation theory, imperfect gases, distribution functions, more. 448pp. 5⅜ × 8½. 65390-0 Pa. $9.95

ORDINARY DIFFERENTIAL EQUATIONS AND STABILITY THEORY: An Introduction, David A. Sánchez. Brief, modern treatment. Linear equation, stability theory for autonomous and nonautonomous systems, etc. 164pp. 5⅜ × 8¼. 63828-6 Pa. $5.95

THIRTY YEARS THAT SHOOK PHYSICS: The Story of Quantum Theory, George Gamow. Lucid, accessible introduction to influential theory of energy and matter. Careful explanations of Dirac's anti-particles, Bohr's model of the atom, much more. 12 plates. Numerous drawings. 240pp. 5⅜ × 8½. 24895-X Pa. $6.95

THEORY OF MATRICES, Sam Perlis. Outstanding text covering rank, non-singularity and inverses in connection with the development of canonical matrices under the relation of equivalence, and without the intervention of determinants. Includes exercises. 237pp. 5⅜ × 8½. 66810-X Pa. $7.95

GREAT EXPERIMENTS IN PHYSICS: Firsthand Accounts from Galileo to Einstein, edited by Morris H. Shamos. 25 crucial discoveries: Newton's laws of motion, Chadwick's study of the neutron, Hertz on electromagnetic waves, more. Original accounts clearly annotated. 370pp. 5⅜ × 8½. 25346-5 Pa. $10.95

INTRODUCTION TO PARTIAL DIFFERENTIAL EQUATIONS WITH APPLICATIONS, E.C. Zachmanoglou and Dale W. Thoe. Essentials of partial differential equations applied to common problems in engineering and the physical sciences. Problems and answers. 416pp. 5⅜ × 8½. 65251-3 Pa. $10.95

BURNHAM'S CELESTIAL HANDBOOK, Robert Burnham, Jr. Thorough guide to the stars beyond our solar system. Exhaustive treatment. Alphabetical by constellation: Andromeda to Cetus in Vol. 1; Chamaeleon to Orion in Vol. 2; and Pavo to Vulpecula in Vol. 3. Hundreds of illustrations. Index in Vol. 3. 2,000pp. 6⅛ × 9¼. 23567-X, 23568-8, 23673-0 Pa., Three-vol. set $41.85

CHEMICAL MAGIC, Leonard A. Ford. Second Edition, Revised by E. Winston Grundmeier. Over 100 unusual stunts demonstrating cold fire, dust explosions, much more. Text explains scientific principles and stresses safety precautions. 128pp. 5⅜ × 8½. 67628-5 Pa. $5.95

AMATEUR ASTRONOMER'S HANDBOOK, J.B. Sidgwick. Timeless, comprehensive coverage of telescopes, mirrors, lenses, mountings, telescope drives, micrometers, spectroscopes, more. 189 illustrations. 576pp. 5⅜ × 8¼. (Available in U.S. only) 24034-7 Pa. $9.95

SPECIAL FUNCTIONS, N.N. Lebedev. Translated by Richard Silverman. Famous Russian work treating more important special functions, with applications to specific problems of physics and engineering. 38 figures. 308pp. 5⅜ × 8½.
60624-4 Pa. $8.95

OBSERVATIONAL ASTRONOMY FOR AMATEURS, J.B. Sidgwick. Mine of useful data for observation of sun, moon, planets, asteroids, aurorae, meteors, comets, variables, binaries, etc. 39 illustrations. 384pp. 5⅜ × 8¼. (Available in U.S. only)
24033-9 Pa. $8.95

INTEGRAL EQUATIONS, F.G. Tricomi. Authoritative, well-written treatment of extremely useful mathematical tool with wide applications. Volterra Equations, Fredholm Equations, much more. Advanced undergraduate to graduate level. Exercises. Bibliography. 238pp. 5⅜ × 8½.
64828-1 Pa. $7.95

POPULAR LECTURES ON MATHEMATICAL LOGIC, Hao Wang. Noted logician's lucid treatment of historical developments, set theory, model theory, recursion theory and constructivism, proof theory, more. 3 appendixes. Bibliography. 1981 edition. ix + 283pp. 5⅜ × 8½.
67632-3 Pa. $8.95

MODERN NONLINEAR EQUATIONS, Thomas L. Saaty. Emphasizes practical solution of problems; covers seven types of equations. ". . . a welcome contribution to the existing literature. . . ."—*Math Reviews.* 490pp. 5⅜ × 8½. 64232-1 Pa. $11.95

FUNDAMENTALS OF ASTRODYNAMICS, Roger Bate et al. Modern approach developed by U.S. Air Force Academy. Designed as a first course. Problems, exercises. Numerous illustrations. 455pp. 5⅜ × 8½.
60061-0 Pa. $9.95

INTRODUCTION TO LINEAR ALGEBRA AND DIFFERENTIAL EQUATIONS, John W. Dettman. Excellent text covers complex numbers, determinants, orthonormal bases, Laplace transforms, much more. Exercises with solutions. Undergraduate level. 416pp. 5⅜ × 8½.
65191-6 Pa. $9.95

INCOMPRESSIBLE AERODYNAMICS, edited by Bryan Thwaites. Covers theoretical and experimental treatment of the uniform flow of air and viscous fluids past two-dimensional aerofoils and three-dimensional wings; many other topics. 654pp. 5⅜ × 8½.
65465-6 Pa. $16.95

INTRODUCTION TO DIFFERENCE EQUATIONS, Samuel Goldberg. Exceptionally clear exposition of important discipline with applications to sociology, psychology, economics. Many illustrative examples; over 250 problems. 260pp. 5⅜ × 8½.
65084-7 Pa. $7.95

LAMINAR BOUNDARY LAYERS, edited by L. Rosenhead. Engineering classic covers steady boundary layers in two- and three-dimensional flow, unsteady boundary layers, stability, observational techniques, much more. 708pp. 5⅜ × 8½.
65646-2 Pa. $18.95

LECTURES ON CLASSICAL DIFFERENTIAL GEOMETRY, Second Edition, Dirk J. Struik. Excellent brief introduction covers curves, theory of surfaces, fundamental equations, geometry on a surface, conformal mapping, other topics. Problems. 240pp. 5⅜ × 8½.
65609-8 Pa. $7.95

ROTARY-WING AERODYNAMICS, W.Z. Stepniewski. Clear, concise text covers aerodynamic phenomena of the rotor and offers guidelines for helicopter performance evaluation. Originally prepared for NASA. 537 figures. 640pp. 6⅛ × 9¼.
64647-5 Pa. $15.95

DIFFERENTIAL GEOMETRY, Heinrich W. Guggenheimer. Local differential geometry as an application of advanced calculus and linear algebra. Curvature, transformation groups, surfaces, more. Exercises. 62 figures. 378pp. 5⅜ × 8½.
63433-7 Pa. $8.95

INTRODUCTION TO SPACE DYNAMICS, William Tyrrell Thomson. Comprehensive, classic introduction to space-flight engineering for advanced undergraduate and graduate students. Includes vector algebra, kinematics, transformation of coordinates. Bibliography. Index. 352pp. 5⅜ × 8½.   65113-4 Pa. $8.95

A SURVEY OF MINIMAL SURFACES, Robert Osserman. Up-to-date, in-depth discussion of the field for advanced students. Corrected and enlarged edition covers new developments. Includes numerous problems. 192pp. 5⅜ × 8½.
64998-9 Pa. $8.95

ANALYTICAL MECHANICS OF GEARS, Earle Buckingham. Indispensable reference for modern gear manufacture covers conjugate gear-tooth action, gear-tooth profiles of various gears, many other topics. 263 figures. 102 tables. 546pp. 5⅜ × 8½.
65712-4 Pa. $14.95

SET THEORY AND LOGIC, Robert R. Stoll. Lucid introduction to unified theory of mathematical concepts. Set theory and logic seen as tools for conceptual understanding of real number system. 496pp. 5⅜ × 8¼.   63829-4 Pa. $10.95

A HISTORY OF MECHANICS, René Dugas. Monumental study of mechanical principles from antiquity to quantum mechanics. Contributions of ancient Greeks, Galileo, Leonardo, Kepler, Lagrange, many others. 671pp. 5⅜ × 8½.
65632-2 Pa. $14.95

FAMOUS PROBLEMS OF GEOMETRY AND HOW TO SOLVE THEM, Benjamin Bold. Squaring the circle, trisecting the angle, duplicating the cube: learn their history, why they are impossible to solve, then solve them yourself. 128pp. 5⅜ × 8½.
24297-8 Pa. $4.95

MECHANICAL VIBRATIONS, J.P. Den Hartog. Classic textbook offers lucid explanations and illustrative models, applying theories of vibrations to a variety of practical industrial engineering problems. Numerous figures. 233 problems, solutions. Appendix. Index. Preface. 436pp. 5⅜ × 8½.   64785-4 Pa. $10.95

CURVATURE AND HOMOLOGY, Samuel I. Goldberg. Thorough treatment of specialized branch of differential geometry. Covers Riemannian manifolds, topology of differentiable manifolds, compact Lie groups, other topics. Exercises. 315pp. 5⅜ × 8½.
64314-X Pa. $8.95

HISTORY OF STRENGTH OF MATERIALS, Stephen P. Timoshenko. Excellent historical survey of the strength of materials with many references to the theories of elasticity and structure. 245 figures. 452pp. 5⅜ × 8½. 61187-6 Pa. $11.95

GEOMETRY OF COMPLEX NUMBERS, Hans Schwerdtfeger. Illuminating, widely praised book on analytic geometry of circles, the Moebius transformation, and two-dimensional non-Euclidean geometries. 200pp. 5⅝ × 8¼.
63830-8 Pa. $8.95

MECHANICS, J.P. Den Hartog. A classic introductory text or refresher. Hundreds of applications and design problems illuminate fundamentals of trusses, loaded beams and cables, etc. 334 answered problems. 462pp. 5⅜ × 8½. 60754-2 Pa. $9.95

TOPOLOGY, John G. Hocking and Gail S. Young. Superb one-year course in classical topology. Topological spaces and functions, point-set topology, much more. Examples and problems. Bibliography. Index. 384pp. 5⅜ × 8¼.
65676-4 Pa. $9.95

STRENGTH OF MATERIALS, J.P. Den Hartog. Full, clear treatment of basic material (tension, torsion, bending, etc.) plus advanced material on engineering methods, applications. 350 answered problems. 323pp. 5⅜ × 8½. 60755-0 Pa. $8.95

ELEMENTARY CONCEPTS OF TOPOLOGY, Paul Alexandroff. Elegant, intuitive approach to topology from set-theoretic topology to Betti groups; how concepts of topology are useful in math and physics. 25 figures. 57pp. 5⅜ × 8½.
60747-X Pa. $3.50

ADVANCED STRENGTH OF MATERIALS, J.P. Den Hartog. Superbly written advanced text covers torsion, rotating disks, membrane stresses in shells, much more. Many problems and answers. 388pp. 5⅜ × 8½. 65407-9 Pa. $9.95

COMPUTABILITY AND UNSOLVABILITY, Martin Davis. Classic graduate-level introduction to theory of computability, usually referred to as theory of recurrent functions. New preface and appendix. 288pp. 5⅜ × 8½. 61471-9 Pa. $7.95

GENERAL CHEMISTRY, Linus Pauling. Revised 3rd edition of classic first-year text by Nobel laureate. Atomic and molecular structure, quantum mechanics, statistical mechanics, thermodynamics correlated with descriptive chemistry. Problems. 992pp. 5⅜ × 8½. 65622-5 Pa. $19.95

AN INTRODUCTION TO MATRICES, SETS AND GROUPS FOR SCIENCE STUDENTS, G. Stephenson. Concise, readable text introduces sets, groups, and most importantly, matrices to undergraduate students of physics, chemistry, and engineering. Problems. 164pp. 5⅜ × 8½. 65077-4 Pa. $6.95

THE HISTORICAL BACKGROUND OF CHEMISTRY, Henry M. Leicester. Evolution of ideas, not individual biography. Concentrates on formulation of a coherent set of chemical laws. 260pp. 5⅜ × 8½. 61053-5 Pa. $6.95

THE PHILOSOPHY OF MATHEMATICS: An Introductory Essay, Stephan Körner. Surveys the views of Plato, Aristotle, Leibniz & Kant concerning propositions and theories of applied and pure mathematics. Introduction. Two appendices. Index. 198pp. 5⅜ × 8½. 25048-2 Pa. $7.95

THE DEVELOPMENT OF MODERN CHEMISTRY, Aaron J. Ihde. Authoritative history of chemistry from ancient Greek theory to 20th-century innovation. Covers major chemists and their discoveries. 209 illustrations. 14 tables. Bibliographies. Indices. Appendices. 851pp. 5⅜ × 8½. 64235-6 Pa. $18.95

DE RE METALLICA, Georgius Agricola. The famous Hoover translation of greatest treatise on technological chemistry, engineering, geology, mining of early modern times (1556). All 289 original woodcuts. 638pp. 6¾ × 11.
60006-8 Pa. $18.95

SOME THEORY OF SAMPLING, William Edwards Deming. Analysis of the problems, theory and design of sampling techniques for social scientists, industrial managers and others who find statistics increasingly important in their work. 61 tables. 90 figures. xvii + 602pp. 5⅜ × 8½.
64684-X Pa. $15.95

THE VARIOUS AND INGENIOUS MACHINES OF AGOSTINO RAMELLI: A Classic Sixteenth-Century Illustrated Treatise on Technology, Agostino Ramelli. One of the most widely known and copied works on machinery in the 16th century. 194 detailed plates of water pumps, grain mills, cranes, more. 608pp. 9 × 12.
25497-6 Clothbd. $34.95

LINEAR PROGRAMMING AND ECONOMIC ANALYSIS, Robert Dorfman, Paul A. Samuelson and Robert M. Solow. First comprehensive treatment of linear programming in standard economic analysis. Game theory, modern welfare economics, Leontief input-output, more. 525pp. 5⅜ × 8½.
65491-5 Pa. $14.95

ELEMENTARY DECISION THEORY, Herman Chernoff and Lincoln E. Moses. Clear introduction to statistics and statistical theory covers data processing, probability and random variables, testing hypotheses, much more. Exercises. 364pp. 5⅜ × 8½.
65218-1 Pa. $9.95

THE COMPLEAT STRATEGYST: Being a Primer on the Theory of Games of Strategy, J.D. Williams. Highly entertaining classic describes, with many illustrated examples, how to select best strategies in conflict situations. Prefaces. Appendices. 268pp. 5⅜ × 8½.
25101-2 Pa. $7.95

MATHEMATICAL METHODS OF OPERATIONS RESEARCH, Thomas L. Saaty. Classic graduate-level text covers historical background, classical methods of forming models, optimization, game theory, probability, queueing theory, much more. Exercises. Bibliography. 448pp. 5⅜ × 8¼.
65703-5 Pa. $12.95

CONSTRUCTIONS AND COMBINATORIAL PROBLEMS IN DESIGN OF EXPERIMENTS, Damaraju Raghavarao. In-depth reference work examines orthogonal Latin squares, incomplete block designs, tactical configuration, partial geometry, much more. Abundant explanations, examples. 416pp. 5⅜ × 8¼.
65685-3 Pa. $10.95

THE ABSOLUTE DIFFERENTIAL CALCULUS (CALCULUS OF TENSORS), Tullio Levi-Civita. Great 20th-century mathematician's classic work on material necessary for mathematical grasp of theory of relativity. 452pp. 5⅜ × 8½.
63401-9 Pa. $9.95

VECTOR AND TENSOR ANALYSIS WITH APPLICATIONS, A.I. Borisenko and I.E. Tarapov. Concise introduction. Worked-out problems, solutions, exercises. 257pp. 5⅜ × 8¼.
63833-2 Pa. $7.95

THE FOUR-COLOR PROBLEM: Assaults and Conquest, Thomas L. Saaty and Paul G. Kainen. Engrossing, comprehensive account of the century-old combinatorial topological problem, its history and solution. Bibliographies. Index. 110 figures. 228pp. 5⅜ × 8½.
65092-8 Pa. $6.95

CATALYSIS IN CHEMISTRY AND ENZYMOLOGY, William P. Jencks. Exceptionally clear coverage of mechanisms for catalysis, forces in aqueous solution, carbonyl- and acyl-group reactions, practical kinetics, more. 864pp. 5⅜ × 8½.
65460-5 Pa. $19.95

PROBABILITY: An Introduction, Samuel Goldberg. Excellent basic text covers set theory, probability theory for finite sample spaces, binomial theorem, much more. 360 problems. Bibliographies. 322pp. 5⅜ × 8½.
65252-1 Pa. $8.95

LIGHTNING, Martin A. Uman. Revised, updated edition of classic work on the physics of lightning. Phenomena, terminology, measurement, photography, spectroscopy, thunder, more. Reviews recent research. Bibliography. Indices. 320pp. 5⅜ × 8¼.
64575-4 Pa. $8.95

PROBABILITY THEORY: A Concise Course, Y.A. Rozanov. Highly readable, self-contained introduction covers combination of events, dependent events, Bernoulli trials, etc. Translation by Richard Silverman. 148pp. 5⅜ × 8¼.
63544-9 Pa. $5.95

AN INTRODUCTION TO HAMILTONIAN OPTICS, H. A. Buchdahl. Detailed account of the Hamiltonian treatment of aberration theory in geometrical optics. Many classes of optical systems defined in terms of the symmetries they possess. Problems with detailed solutions. 1970 edition. xv + 360pp. 5⅜ × 8½.
67597-1 Pa. $10.95

STATISTICS MANUAL, Edwin L. Crow, et al. Comprehensive, practical collection of classical and modern methods prepared by U.S. Naval Ordnance Test Station. Stress on use. Basics of statistics assumed. 288pp. 5⅜ × 8½.
60599-X Pa. $6.95

DICTIONARY/OUTLINE OF BASIC STATISTICS, John E. Freund and Frank J. Williams. A clear concise dictionary of over 1,000 statistical terms and an outline of statistical formulas covering probability, nonparametric tests, much more. 208pp. 5⅜ × 8½.
66796-0 Pa. $6.95

STATISTICAL METHOD FROM THE VIEWPOINT OF QUALITY CONTROL, Walter A. Shewhart. Important text explains regulation of variables, uses of statistical control to achieve quality control in industry, agriculture, other areas. 192pp. 5⅜ × 8½.
65232-7 Pa. $7.95

THE INTERPRETATION OF GEOLOGICAL PHASE DIAGRAMS, Ernest G. Ehlers. Clear, concise text emphasizes diagrams of systems under fluid or containing pressure; also coverage of complex binary systems, hydrothermal melting, more. 288pp. 6½ × 9¼.
65389-7 Pa. $10.95

STATISTICAL ADJUSTMENT OF DATA, W. Edwards Deming. Introduction to basic concepts of statistics, curve fitting, least squares solution, conditions without parameter, conditions containing parameters. 26 exercises worked out. 271pp. 5⅜ × 8½.
64685-8 Pa. $8.95

TENSOR CALCULUS, J.L. Synge and A. Schild. Widely used introductory text covers spaces and tensors, basic operations in Riemannian space, non-Riemannian spaces, etc. 324pp. 5⅜ × 8¼. 63612-7 Pa. $8.95

A CONCISE HISTORY OF MATHEMATICS, Dirk J. Struik. The best brief history of mathematics. Stresses origins and covers every major figure from ancient Near East to 19th century. 41 illustrations. 195pp. 5⅜ × 8½. 60255-9 Pa. $7.95

A SHORT ACCOUNT OF THE HISTORY OF MATHEMATICS, W.W. Rouse Ball. One of clearest, most authoritative surveys from the Egyptians and Phoenicians through 19th-century figures such as Grassman, Galois, Riemann. Fourth edition. 522pp. 5⅜ × 8½. 20630-0 Pa. $10.95

HISTORY OF MATHEMATICS, David E. Smith. Nontechnical survey from ancient Greece and Orient to late 19th century; evolution of arithmetic, geometry, trigonometry, calculating devices, algebra, the calculus. 362 illustrations. 1,355pp. 5⅜ × 8½. 20429-4, 20430-8 Pa., Two-vol. set $23.90

THE GEOMETRY OF RENÉ DESCARTES, René Descartes. The great work founded analytical geometry. Original French text, Descartes' own diagrams, together with definitive Smith-Latham translation. 244pp. 5⅜ × 8½. 60068-8 Pa. $6.95

THE ORIGINS OF THE INFINITESIMAL CALCULUS, Margaret E. Baron. Only fully detailed and documented account of crucial discipline: origins; development by Galileo, Kepler, Cavalieri; contributions of Newton, Leibniz, more. 304pp. 5⅜ × 8½. (Available in U.S. and Canada only) 65371-4 Pa. $9.95

THE HISTORY OF THE CALCULUS AND ITS CONCEPTUAL DEVELOPMENT, Carl B. Boyer. Origins in antiquity, medieval contributions, work of Newton, Leibniz, rigorous formulation. Treatment is verbal. 346pp. 5⅜ × 8½. 60509-4 Pa. $8.95

THE THIRTEEN BOOKS OF EUCLID'S ELEMENTS, translated with introduction and commentary by Sir Thomas L. Heath. Definitive edition. Textual and linguistic notes, mathematical analysis. 2,500 years of critical commentary. Not abridged. 1,414pp. 5⅜ × 8½. 60088-2, 60089-0, 60090-4 Pa., Three-vol. set $29.85

GAMES AND DECISIONS: Introduction and Critical Survey, R. Duncan Luce and Howard Raiffa. Superb nontechnical introduction to game theory, primarily applied to social sciences. Utility theory, zero-sum games, n-person games, decision-making, much more. Bibliography. 509pp. 5⅜ × 8½. 65943-7 Pa. $12.95

THE HISTORICAL ROOTS OF ELEMENTARY MATHEMATICS, Lucas N.H. Bunt, Phillip S. Jones, and Jack D. Bedient. Fundamental underpinnings of modern arithmetic, algebra, geometry and number systems derived from ancient civilizations. 320pp. 5⅜ × 8½. 25563-8 Pa. $8.95

CALCULUS REFRESHER FOR TECHNICAL PEOPLE, A. Albert Klaf. Covers important aspects of integral and differential calculus via 756 questions. 566 problems, most answered. 431pp. 5⅜ × 8½. 20370-0 Pa. $8.95

CHALLENGING MATHEMATICAL PROBLEMS WITH ELEMENTARY SOLUTIONS, A.M. Yaglom and I.M. Yaglom. Over 170 challenging problems on probability theory, combinatorial analysis, points and lines, topology, convex polygons, many other topics. Solutions. Total of 445pp. 5⅜ × 8½. Two-vol. set.

Vol. I 65536-9 Pa. $7.95
Vol. II 65537-7 Pa. $6.95

FIFTY CHALLENGING PROBLEMS IN PROBABILITY WITH SOLUTIONS, Frederick Mosteller. Remarkable puzzlers, graded in difficulty, illustrate elementary and advanced aspects of probability. Detailed solutions. 88pp. 5⅜ × 8½.

65355-2 Pa. $4.95

EXPERIMENTS IN TOPOLOGY, Stephen Barr. Classic, lively explanation of one of the byways of mathematics. Klein bottles, Moebius strips, projective planes, map coloring, problem of the Koenigsberg bridges, much more, described with clarity and wit. 43 figures. 210pp. 5⅜ × 8½. 25933-1 Pa. $5.95

RELATIVITY IN ILLUSTRATIONS, Jacob T. Schwartz. Clear nontechnical treatment makes relativity more accessible than ever before. Over 60 drawings illustrate concepts more clearly than text alone. Only high school geometry needed. Bibliography. 128pp. 6⅛ × 9¼. 25965-X Pa. $6.95

AN INTRODUCTION TO ORDINARY DIFFERENTIAL EQUATIONS, Earl A. Coddington. A thorough and systematic first course in elementary differential equations for undergraduates in mathematics and science, with many exercises and problems (with answers). Index. 304pp. 5⅜ × 8½. 65942-9 Pa. $8.95

FOURIER SERIES AND ORTHOGONAL FUNCTIONS, Harry F. Davis. An incisive text combining theory and practical example to introduce Fourier series, orthogonal functions and applications of the Fourier method to boundary-value problems. 570 exercises. Answers and notes. 416pp. 5⅜ × 8½. 65973-9 Pa. $9.95

THE THEORY OF BRANCHING PROCESSES, Theodore E. Harris. First systematic, comprehensive treatment of branching (i.e. multiplicative) processes and their applications. Galton-Watson model, Markov branching processes, electron-photon cascade, many other topics. Rigorous proofs. Bibliography. 240pp. 5⅜ × 8½. 65952-6 Pa. $6.95

AN INTRODUCTION TO ALGEBRAIC STRUCTURES, Joseph Landin. Superb self-contained text covers "abstract algebra": sets and numbers, theory of groups, theory of rings, much more. Numerous well-chosen examples, exercises. 247pp. 5⅜ × 8½. 65940-2 Pa. $7.95

Prices subject to change without notice.
Available at your book dealer or write for free Mathematics and Science Catalog to Dept. GI, Dover Publications, Inc., 31 East 2nd St., Mineola, N.Y. 11501. Dover publishes more than 175 books each year on science, elementary and advanced mathematics, biology, music, art, literature, history, social sciences and other areas.